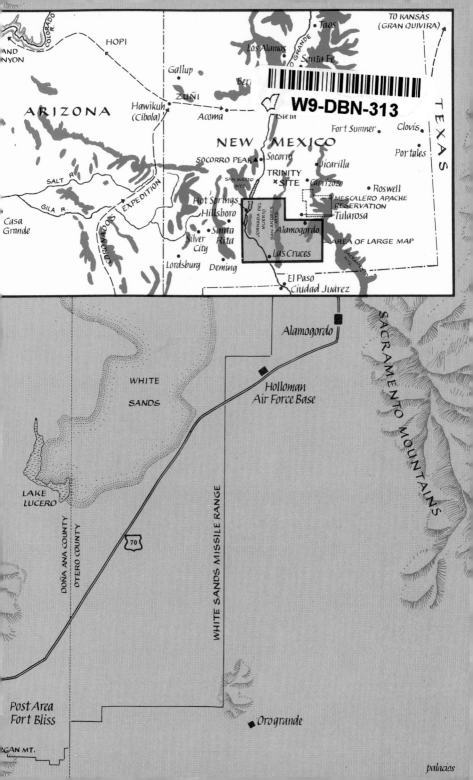

W9-DBN-313

ONE HUNDRED
TONS OF GOLD

Also by David Leon Chandler

BROTHERS IN BLOOD
THE NATURAL SUPERIORITY OF SOUTHERN POLITICIANS

ONE HUNDRED

by

1978 GARDEN CITY, NEW YORK

TONS OF GOLD

David Leon Chandler

DOUBLEDAY & COMPANY, INC.

PHOTO CREDITS
Courtesy of Norm Bergsma: 1, 2, 3, 4, 5, 6, 7, 8, 18, 19, 20, 21
Courtesy of the El Paso *Herald Post:* 9
Courtesy of Joe Newman: 11
Courtesy of U. S. Army: 13, 14, 15

Library of Congress Cataloging in Publication Data
Chandler, David Leon.
    One hundred tons of gold.

    Includes index.
    1. Treasure-trove—New Mexico—Victorio Peak.
2. Victorio Peak, N.M. 3. Gold. I. Title.
F802.S15C46   978.9'66
ISBN: 0-385-12738-3
Library of Congress Catalog Card Number 77–73327

to S.O.

My thanks and grateful acknowledgment for assistance in preparing this manuscript to Howard Bryan of the Albuquerque *Tribune;* Joe and Joan Newman; Kevin Henry; Norman Scott; F. Lee Bailey and Wayne Smith; Charles D. Ablard, general counsel, Department of the Army; Howard Hinton, editor of *Arizona West;* Stewart Peckham, Museum of New Mexico; Myra Ellen Jenkins, New Mexico state historian; Robert Weber, senior geologist, New Mexico Bureau of Mines; Oscar Jordan, counsel, state land office; and Eve Ball, the historian on the Apaches.

Contents

Me experyence with goold minin'
is it's always in the nex' county.
        —*Finley Peter Dunne's*
         "Mister Dooley"

ONE HUNDRED
TONS OF GOLD

This story is true. Its source materials are interviews and private correspondence; documents of the New Mexico state land office, Bureau of Mines, and Museum of New Mexico; and previously unreleased federal records obtained through the Freedom of Information Act from the Department of the Army, Secretary of the Treasury, United States Mint, Office of Domestic Gold and Silver Operations, Securities Exchange Commission, and the Secret Service.

WHITE HOUSE GOLD

Egil Krogh had described to me how, when he was bored with his desktop, he had carried bars of gold bullion through Asia's "Golden Triangle" in CIA planes and bargained with drug chieftains.

—White House counsel John Dean,
Blind Ambition (*New York:*
Simon & Schuster, 1976), *p. 155*

WHITE SANDS GOLD

This operation has been carried as a Top Secret project; and the only persons at White Sands Missile Range who know the nature of the mission are [names censored].

—Communication of the Secret Service
(*Aug. 10, 1961*)

ARMY GOLD

The 90th Division had stumbled into an underground cache containing the Reich's last gold reserves. . . . In addition to $100 million in gold bullion, the MPs found 3 billion Reichsmarks. Another $2 million in American greenbacks together with lesser quantities of British, Norwegian, and French currency had been stacked in those dry salt chambers 2,100 feet below the ground. Eisenhower and I went down in the double-tiered lift with a German workman operating the hoist. The bullion, in 25-pound bars, was packed two to a sack and stenciled in black with the imprint of the Reichsbank. . . .

In a stack nearby we examined hundreds of crates and boxes, art treasures that had been removed from Berlin for safekeeping.

We joked with Patton on the discovery. "If these were the old freebooting days," I said, "when a soldier kept his loot, you'd be the richest man in the world." Patton only grinned back. . . .

Patton had ordered a censorship on the discovery. . . .

"Why keep it a secret, George?" I laughed. "What would you do with all that money?"

George chuckled. Third Army was divided into two schools on the issue, he said. One recommended that the gold be cut up into medallions. "One for every sonofabitch in Third Army—" The other suggested Third Army hide the loot until peacetime when Congress again cracked down military appropriations. Then when-

ever funds got tight, the army could dig down into its cave for more money to spend on new weapons.

Ike shook his head, looked at me, and laughed. "He's always got an answer," he said.

—*General Omar Bradley,* A Soldier's Story
*(New York: Henry Holt and Company,
1951), pp. 540–41*

I

Noss's Gold

In the southeastern New Mexico desert there rises a range of mountains called the San Andres. Bleak, nearly waterless, they are covered by meager growth of mesquite, prickly pear, and piñon tree. The San Andres were once the mountain home of the Mescalero Apache, the launching site for the Apaches' 350-year war with the Spanish, the Mexican, and the American invaders of New Mexico.

The San Andres are isolated, without population, and distant from civilization. At the northern end of the range is a pit of melted rock and glass. It is Trinity Site where the United States exploded the world's first atomic bomb in 1945. Today the San Andres are part of the White Sands Missile Range. The land, once lightly touched by Apache moccasins, now trembles with the roar of Cruise missiles whooshing through the twisting valleys, missiles that rise and dip with the terrain at one hundred feet elevation.

It is said that the San Andres contain a fabulous treasure, one hundred tons of gold, wealth enough to topple governments. Rumors of this treasure surfaced in the 1973 Watergate hearings held by the United States Senate.

The treasure is said to lie in Victorio Peak, a five-hundred-foot-high basalt cone named after an Apache war chief. The cone sits in the center of a huge natural amphitheater called Hembrillo Basin, a nearly circular saucer with rims of six thousand feet elevation.

At dawn on Saturday, March 19, 1977, the testing of missiles was suspended. A caravan of nearly seventy vehicles gathered at the west end of a thirty-mile desert that leads to Victorio Peak. Present at the gathering were adventurers, professional treasure hunters, archaeologists, geologists, two physicists, and ninety-eight national and international journalists.

They were looking for the gold. Four years after Watergate, after countless negotiations, the Army had agreed to the treasure hunt. It was code-named "Goldfinder."

Treasure-troves that have been lost for centuries are found now and then. Gold coins, twelve hundred of them dating from the reign of Richard II, were found by an Englishman in 1966. In the same year, an estimated $3.5 million in bullion was recovered from the wreck of a Spanish fleet which had crashed and sunk off the Florida coast. In 1968, a group of Peruvian treasure hunters found an estimated $10 million in Inca gold.

An even more successful venture was documented by the *National Geographic,* which in 1976 released a film on Florida diver Mel Fisher's five-year quest to find the Spanish galleon *Atocha.* The *Atocha* had sunk in 1622 near the Florida keys with a cargo that included hundreds of bars of gold and silver, 260,000 silver coins, and 15 tons of Cuban copper, plus the personal effects of its wealthy passengers. Fisher found the *Atocha* and recovered large portions of a total treasure estimated at $100 million in value.

In October 1976, a find of possibly equal value was reported near Vera Cruz, Mexico. A twenty-six-year-old fisherman, Raul Hurtado, discovered between fifty and sixty gold bars and finely wrought ornaments in a shallow cove about two miles north of Vera Cruz.

Alfonso Medellin Zenil, director of Vera Cruz University's anthropology institute, said that the worked ornaments dated back to about A.D. 1200. He suggested that the bars, ornaments, and jewelry may have been part of the Aztec Emperor Montezuma's treasure, which was known to have been looted by Cortez and carried back to Vera Cruz for shipment to Spain.

The *Atocha* and *Vera Cruz* treasures are the most valuable among known treasures. But the Victorio Peak trove, if it exists, compares to them as Mount Everest to a pile of small change.

It is the biggest of treasures and the most mysterious. History recorded the sinking of the *Atocha* and the looting of the Aztec

treasure house. The Victorio Peak treasure has no such creden-
tials. But it has had a lot of action.

Rumors of a vast treasure in the region of Hembrillo go back
for more than a century. Official records of the rumors do not
begin, however, until November 18, 1929.

On that day a man identifying himself as Willie Daught walked
into the Las Cruces, New Mexico, sheriff's office and asked for
protection. Daught claimed that twelve days earlier he had found
gold bars in the Caballos Mountains, a small range running parallel to the San Andres.

The bars, said Daught, were stacked in a single pile ten feet
long and five feet high.

He had brought a bar out and, after showing it to some friends,
took it with him to Hatch, the only town in the region. There he
made arrangements to sell the bars.

At around midnight, three armed men broke into his room.
With drawn revolvers, they demanded to know the location of
Daught's treasure. He refused to tell them. They led him down-
stairs to a car and drove him back up into the Caballos. There
they made camp. Daught was chained to a tree. He was beaten
and tortured with lighted cigarettes. After three days of this, the
kidnapers decided to move their camp, and in the process Daught
managed to escape.

He headed first for the hotel in Hatch to see if the gold bar was
still in the safe. It was. Daught then went to the sheriff in Las
Cruces and asked for protection.[1]

About two days later, according to persistent gossip, a convoy
of Army trucks arrived in Hatch, then headed into the Caballos,
where they loaded Daught's gold, or someone's gold, and took it
East.

Daught himself remained in the region for another two years.
By all accounts available today, he spent the time moving and
selling gold. In 1933, he disappeared from New Mexico.

Thirty years later, although his name was forgotten, he and his
treasure had become legend, literally.

California sociologist Ruth Tolman, researching the "myths of

[1] El Paso *Herald,* Nov. 19, 1929.

lost treasure," visited the Hatch region numerous times during the
period 1947–60. There in thirty-two interviews she found versions
of a persistent myth. Four features were common to the interviews: the Caballos Mountains; caves; fresh-water springs in the
caves; and gold.

Bud H., aged seventy-three, interviewed in 1950, said in 1928
that he grubstaked a boy who had a treasure chart and who spent
three months looking around near Hatch. "Late one evening he
stumbled into my place just exhausted . . . he showed me two
bars of heavy metal . . . he told me something about the place
where he had found the stuff. It was in a cave, and there were old
guns there and a bellows made of rawhide. He had crawled down
through a narrow tunnel. . . . The bars were stacked like cordwood in several different stacks. . . . There was a human skeleton lying on one of the stacks. . . . I took a hacksaw and sawed
off a little piece. . . . Those people at the assay office acted like
crazy folks: They asked me, 'Do you know this is gold?' . . . The
boy went back for more bars. I guess it was five or six days after
that these other two fellows showed up with some of the same
stuff, the same gold. The boy never came back. . . . But word
went around that gold had been found. . . . Others come in—
from Oklahoma, California, and from everywhere . . . dozens of
men prowling those mountains, all wearing guns and not trusting
each other. . . . But they never did find it. Nobody has found it
yet."

C. R. Buckelew, aged sixty-one, was interviewed at Hatch in
1959: "One evening sometime in 1928 I was getting ready to
close my service station. A fellow drove up and wanted gasoline.
He was in an awful hurry and said, 'Charlie, I need gas to get to
El Paso. I haven't got any money now, but I've got something in
the back of the car here that will make us rich. I found some bars
of solid gold up in the mountains. . . . You have to let me have
some gas, and I'll sure make it right with you.' He was in a hurry,
and I didn't ask to see the bars. I let him have the gas; I would
have anyway. I heard later that he was trying to get away from
someone who was following him and tossed the gold bars in an
empty freight car that was moving. He slipped as he tried to jump
in after them, fell under the train and it cut off both his legs."

Lester Buckelew, aged forty-three, was interviewed in 1960. "Two fellows found some solid gold bars in a cave in the Caballos in the late twenties or early thirties. One stayed at the Jornada Hotel in Hatch that night. Someone found out and was after him. Trying to get away, he fell under a train and both his legs were cut off. Those are the two bars that wound up in a bank in El Paso."

And Gene W., interviewed in 1947, said: "Right after the gold was found in 1928, the government came in with cars, three trucks, and some soldiers from Fort Bliss to load all that gold up and take it back with them. They followed this fellow around over the hills . . . after leading them around all day, the man sat down on a rock and cried like a baby and said, 'I can't do it. I don't know where it is.'"

Ruth Tolman collected four other interviews along the same line. Each of the four said that in 1928 two gold bars had been taken from a cave where gold was "stacked like cordwood." In all the stories the mine is lost and never found again. Descriptions of the cave in each instance place it within twenty miles of Hembrillo Basin.

Finally, in a 1959 interview with an aged Mexican, Ruth Tolman hit upon a possible source for the gold:

"In the early days," he said, "of the Spanish settlement there was a missionary who lived near the present location of the village of Las Palomas. He had his people mining gold . . . but he saved the gold to use for his people instead of sending it to Spain. The Spaniards sent soldiers to collect the funds, but the missionary was warned. . . . He and his people took the gold and all the church valuables high in the mountains. . . . They buried these things in three graves on the side of the mountain. . . ."

The first set of interviews refers to Willie Daught. It is unexplained why the Hatch people all agreed that the gold find occurred in 1928 instead of 1929, as shown by the sheriff's office records. But the other elements hold up—for instance, the train accident. As we will see later, Daught had a friend helping him move the gold. The boy's name was Buster Ward, and Buster Ward would lose both legs trying to throw gold bars into a train.

Willie Daught disappeared intentionally, for mysterious and possibly sinister reasons, from New Mexico in 1933. He said at the time that he would return when it was legal to own gold again.[2] Daught went to California, where he lived comfortably, although not lavishly, under an assumed name. He returned briefly to Hatch in 1976 when, indeed, it was legal to own gold again.

Daught was gone two years when his sequel arrived in Hatch. The year was 1935, and the new treasure hunter was Milton "Doc" Noss, fresh out of the New Mexico state penitentiary.

Doc Noss was thirty years old when he arrived in Hatch. The town looked then much as it does now, its population hovering between eight hundred and nine hundred people with one main street lined with one- and two-story buildings of brick or wood frame or adobe. Hatch's economic role, from the time it was founded in 1881 to the present, is to serve as a market and railroad stop for the surrounding Rio Grande Valley farms and cattle ranches. The hotels that were downtown in Doc Noss's day no longer operate. But the buildings are still there—either vacant or converted to other uses. The town's single downtown movie has been long closed. Otherwise, the town has not much changed. There are a few more paved streets, more cars, and fewer of the Art Deco neon signs that advertise bars and cafes.

The cafes still serve steaks cooked "ranch style"—which is to say that they arrive on the plate tough and burned. You can get a large steak-and-eggs dinner, however, for less than four dollars. The meat comes from local range cattle and probably doesn't pass through the hands of any middleman, including the Department of Agriculture inspector.

Doc Noss was a half-blood Cheyenne Indian from Oklahoma. In 1931, when he was twenty-six years old, he married a woman named Ova, ten years his senior. With her, he wandered the Southwest practicing without a license as a "foot doctor." In 1934, he was arrested in Wellington, Texas, on a charge of practicing without a license. A friend bailed him out, and Doc and Ova split town, jumping bond. They went to Santa Fe, New Mexico, where somehow Doc obtained a commission as deputy sheriff. The post carried no salary, but it allowed him to carry a gun. In 1935, because of the gun, he did four months hard time in the

[2] Ownership of gold by American citizens became illegal after Franklin Roosevelt signed the law on June 5, 1933. It was made legal again in 1975.

New Mexico state prison. The charge was threatening a waitress, in Roswell, while armed.

Her name was Jo Ann. "I drank four glasses of beer and was under the influence of the glasses," Doc wrote in his parole application. "I do not remember what I said to Jo Ann after I drank the glasses in the beer garden." He had been sentenced to serve six to nine months. He got out in four. He would have no further arrests in his lifetime. He was never charged with being a swindler or operating confidence games.

He was a tall, dark-haired, handsome, and eye-catching figure. Joe Andreg, who worked for him in the late 1930s, remembered: "He wore a fancy, all-black western outfit and carried a gun. . . . I was sixteen years old and greatly admired him."

When Doc left prison, Ova and he rented a frame house near Hatch. The house was about thirty-five miles southwest of Hembrillo Basin, which was then state property and accessible to ranchers, deer hunters, Indians, and anyone who had a notion to go there.

Here some geography is helpful.

Hatch and its immediately adjacent neighbor, Rincon, sit astride the Rio Grande River, Hatch on the west bank, Rincon on the east. Running due north from the two towns for about thirty miles is the Caballo Mountain Range. There Willie Daught supposedly found his gold bars, and it is the site referred to by the old Mexican in his tale of the padre's buried treasure.

To the immediate east of the Caballos is a higher and parallel range. This is the San Andres. Hembrillo Basin is part of the San Andres Range.

The San Andres range runs north and south, and Hembrillo is halfway up the range. Hembrillo, a craggy rim surrounding a desert floor, looks like a dormant volcano, which it is not. On the floor of the basin rise two conelike peaks. One of these is Victorio. The other, a twin brother one mile to the north, is Geronimo.

Immediately west of the San Andres is La Jornada del Muerto, "The Dead Man's Route," once part of the Spanish road between Mexico City and Santa Fe. It is a cheerless tableland running north and south between the San Andres and the Rio Grande. It is forty-five miles wide by ninety miles long, with desolate mountains on each side, a lava bed on the north, and the Rio Grande arcing across the south.

This desert was once part of the Spanish caravan route known as the Camino Real to the Spanish and the Santa Fe Trail to Americans. It was preferred by caravans because it was a shorter route than the Rio Grande Valley, and the Jornada's hard-baked red sand gave better support for wagon wheels than the valley's soft loam.

Hundreds, perhaps thousands, of men, women, and children have died on the Jornada. They died from heat, they died from enemies, they died from lack of water and, ironically, they died from drowning. Drowning is an ever-present possibility in the Jornada.

The Jornada is cut laterally by thousands of riverbeds called arroyos. Bone-dry and rock-hard most of the year, periodically the arroyos fill to the brim and spill over with water rushing down from storms in the mountains.

There are only a few seconds' warning. One moment the arroyo is a bed of dry, scrubbed rock. The next moment there is a roaring, followed instantly by a wall of water four feet or five feet high and sometimes higher. It bowls over and crushes anything in its path: people, animals, wagons, automobiles. There is no season. It comes winter and summer, spring and fall. It always comes without warning.

The flash floods and the other elements have caused many deaths on the Jornada. There is no record, not even a close estimate of the number of deaths. But the main killer of people, the main reason for the Jornada's name, is the Apache. It was his primary raiding ground against the Spanish, the Mexican, and the American. It was his battleground for 350 consecutive years, scene of an Apache war that began in the middle of the sixteenth century and ended only at the close of the nineteenth. A bare thirty-five years later, Doc Noss came to the Jornada.

The house of Ova and Doc Noss, Ova remembers, "had a dirt yard, an outdoor john, and water from the pump out back." One November day in 1937, "a bunch of us were going to get meat for the table, and we loaded up in pickups to find some deer in the San Andres."

Traveling across the Jornada, the party headed for Hembrillo Basin. Hembrillo contains a fresh-water spring at the base of Victorio Peak, and there they figured to find deer. They camped in the basin, filled their canteens with the cold, clear water, and set out for deer.

"Doc became separated from the rest of us," Ova later recalled. "After he was missing for a while we began to worry that he might have fallen and broken a leg."

Toward sunset, Doc walked into camp. He took Ova aside and said, "Get your work done and come to bed—I have something to tell you." He had found the shaft.

"Ova, we're going to come back later, when nobody's around."

They returned several days later. Ova remembers that "Doc squeezed down under the edge of the flat rock and went down the pole. A rung broke and he came up, got some lariats, and lowered himself back down. He was gone a couple of hours."

According to papers later filed with the New Mexico state land office, which administered the Hembrillo Basin at the time, Doc said he descended by rope for sixty feet and found himself in a large room. He played the flashlight around and saw "Indian drawings" painted and carved into the stone walls. Sitting on a large rock, studying the walls, Doc heard sand and pebbles dropping into what sounded like a space below. He pushed aside the rock and saw another shaft, descending at an angle. The shaft sloped downward for 126 feet, leveled, then ascended. "It broke into a large cave," Noss told a state investigator, "big enough for a freight train and leading into small cave rooms along the side." An ice-cold stream ran along the floor of the cavern, and Doc estimated that the complex of caves stretched for more than half a mile.

Stumbling along the path of the stream, Doc made his way from room to room. The only sounds were the eerie, echoing flow of the stream and his own breath rasping in his throat. The flashlight beam shone dully off the walls, then settled shaking, upon a grisly find. Doc steadied himself against the cavern wall. There, beside the river, was a human skeleton. "He had red hair and was grinning up at me," said Doc.

The skeleton had collapsed from a kneeling position, and the hands were trussed behind the back with a rotting rope. Doc claimed that further exploration revealed a row of twenty-six more skeletons.

In one of the smaller rooms, just beyond the skeletons, Doc found what he had half expected: treasure. He saw old Wells Fargo chests, swords, guns, saddles, jewels, boxes full of old letters, and "enough gold and silver coins to load sixty to eighty mules." Doc filled his pockets with coins and jewels and started happily out of the cave. Then—back near the shaft, in a corner of the main cave and covered with old buffalo hides—he found thousands upon thousands of bars of pig iron, "stacked like cordwood."

Back on the surface, he showed Ova the gold coins and the jewels. She said he should have brought up one of the pig-iron bars. When he did, she rubbed it clean and said, "Why, Doc! It's *gold!*"

For Ova, it was probably the "happiest moment in our lives. Right then, when he first came up, we knew we had something. We knew it was ours."

Gold. They had hit the greatest strike in history, all of it mined and stacked in thousands and thousands of bars. It was like the impossible things one reads about in books but never sees in life. Gold in such amounts it could influence governments, buy banks, corner markets, crush competitors, purchase public office, accrue vast duchies of land, gold that, if manipulated with ruthless cunning and murderous skill, could gain a nation.

It is ironic that if they had possessed monstrous ambitions and evil talents, they would not have lost the gold. Had their vices been large enough, they would not have lost the gold. Had they possessed an irresistible instinct for power, they would have managed, with scarcely any trouble, to take the gold from the ground and implant it into commerce and government to bloom and seed. Unfortunately for them, those dimensions were not in their characters.

This money they loved with an innocent affection. They didn't love it for itself alone but as wings of flight—a deliverance from the brutish life of being poor. Gold was in their hands, and they were astonished, happy, vain—a vanity of limited imagination. He would be the No. 1 man in all the West, she the leading lady. Fine cars and a large ranch in Oklahoma and college for children and maybe even the governor would come to call. All these things Ova imagined. My! Wouldn't heads turn?

They got to their knees and prayed. "Our Father . . . lead us not into temptation . . ." But it was too late. Gold had worked its change and made them already narrow and suspicious. When they stood up from their prayer, she learned that he wouldn't share.

"You will never go down in the cave," said Doc darkly.

Why?

He looked at her intently. "You might get hurt."

In the months ahead, while she stood guard with a rifle, he alone descended. He brought bars up and would send her away so he could hide the gold around the mountain in places secret to her.

I worked with Doc Noss to carry gold out of the mine. In 1939 I took one bullion bar to Douglas, Arizona, and had Holly and Holly assayers test it, and it assayed to run over $5,000 gold per ton. . . . The bar was a regular Wells Fargo gold bar. I saw quite a few other bars of the same type.

—B. D. Lampros, in an affidavit submitted in October 1952 to the Secretary of Defense

According to Ova Noss, according to the Lampros affidavit, and according to similar affidavits signed by four other men who saw the gold, between 1937 and 1939 Doc Noss pulled out eighty-eight gold bars weighing forty to eighty pounds each and cached them in the basin around the peak. In 1939 the treasure was lost when Noss and an engineer named S. E. Montgomery attempted to enlarge the entrance and instead caused a slide that sealed the cave. This was witnessed by Benny Samaniego, a worker hired by Doc. Samaniego was interviewed during a New Mexico state museum investigation in 1963.

"I was there that day," said Samaniego, "and Doc and Montgomery were arguing. Doc thought it was a bad idea, but Montgomery said he knew what he was doing. . . . I guess he didn't, because he blew it up."

Samaniego and a man named José Serafin Sedillo are the only

persons other than Doc known to have entered the main cave. Both were interviewed during the state museum investigation in 1963.

"I saw stacks of gold bars, skeletons, armor, old guns, and statues," said Samaniego, who worked for Doc from 1937 to 1939. "The skeletons were tied, kneeling to posts, as if they were prisoners left to die."

Sedillo, then a boy living in Rincon, was hired right after Doc's initial discovery. Sedillo made only one trip to the treasure room: "We went in the afternoon, because Doc had learned that at 2 P.M. the sun showed through a hole in the roof of the main cave. . . . The gold bars were stacked up like cordwood. . . . I grabbed two small ones and stuffed them into my shirt. . . . I had a hell of a time climbing up the rope with those things." When he got to the surface he told Doc he was going to keep the gold.

"Doc said, 'No, you aren't.' He took out his gun and said he would kill me first. I gave him the gold and left. . . . The next time I come around there, a couple of years later, the cave was blasted shut by somebody named Montgomery."

In 1949, Doc was shot and killed by another gold hunter.

In 1961, Ova Noss caught the Army mining the mountain in a Top Secret operation.

And in 1977, Ova Noss, eighty years old, climbed to the very apex of Victorio Peak and stood on the top as an uninvited, unwanted claimant in "Operation Goldfinder." A volcano of a woman, alive and near eruption, she stood on a ledge that overlooked the entire basin. Broad and stout, leaning on a strong walking staff, her white hair flowing in the wind, she looked like a desert prophet—and, prophetlike, she discharged her contempt for the Army in words that struck like thunderbolts.

"The goddamn Army has stole the gold," she shouted to the basin. "They have dug it out and hauled it off, the sonofabitches!"

The gold had been there.

"Doc took five bars of it," she said later, "to the U. S. Mint in Denver, where they drilled holes in them and told Doc they were worth $20.67 per troy ounce. She was precise on the Mint's assay figure: *$20.67 per troy ounce.* She cited the figure in a statement to the New Mexico land office in 1952 and has used it since:

$20.67 per troy ounce. It is important. It partially supports
Doc's claim.

Ova said the gold was assayed by the Denver Mint in 1938. The
price of gold then was $35 an ounce—for pure 24-karat gold. In-
directly, Ova is saying that Doc's gold assayed at 60 per cent, 15
karats. In 1973, lawyer F. Lee Bailey, representing clients in "Operation Goldfinder," would present gold to the U. S. Treasury in
Washington for appraisal. It assayed at 61 per cent. In 1974,
Bailey took another bar to Los Alamos for assay. It came out at
63 per cent. The similarities suggest that Noss and Bailey may
have been handling gold from the same smelter.

Ova accused the Army of stealing the gold. Later, F. Lee Bailey
would accuse the Army of the same thing.

Ova's first complaint came in October 1961, when she filed
affidavits with the state land office. She claimed that the Army was
illegally mining her claim; illegally because—regardless of the
gold being on a federal reservation—neither it nor any other mineral can be mined or troved in New Mexico without clearance
from the state land office.

The Army spokesman at White Sands officially denied Ova's ac-
cusation. He denied it to her, he denied it to her lawyer, and he
denied it to the state land office.

Ova then presented more affidavits. These were signed by civilians who had sneaked into Hembrillo and had seen light generators, Jeeps, radios, mining wedges, and timbers. Light cables had
been run into the caves of the old Noss claim. A captain ran the
civilians off, saying the area was restricted. But before they left,
the civilians got license numbers of the military vehicles. And they
got a few names of the personnel doing the mining. All this was in
the affidavits.

It was only then, caught, that the Army owned up to what it
was doing: mining the Noss claim. It was mining the claim, it
turned out, with the full authorization of the Secretary of the
Army.

It seemed that three years earlier, in 1958, an Air Force captain, Leonard Fiege, and an airman, Thomas Berchett, were in the
Hembrillo Basin looking for the Noss treasure at Victorio Peak.

In an affidavit, Fiege said he "went down a canyon by myself. I saw this small hill and open caves. . . . I decided to look around."

He crawled through a series of caves and entered a main cavern. "The dust was so thick and the air was foul and hard to breathe, so I sat down on what I thought was a dust-covered pile of rocks. . . . I started inspecting this area. The pile of so-called rocks was not rocks but smelted gold in bars about the size of a house brick. . . . With my flashlight I saw three piles of this gold all lined up and another pile off to the left that was partly covered by the wall of the cave that had fallen in. . . . I found it impossible to see farther than about fifteen to twenty feet because of the dust in the air."

Fiege got sick and left the cave. He returned to the site with other hunters—Ken Prather, Milt Wessel, and Berchett. Fiege and Berchett re-entered the cave. "We went to those four piles of gold and confirmed the findings." They both became sick and went outside, where the four "decided against taking some of it out because we were not familiar with laws that governed the claiming of this gold. We might lose it all. . . . We decided the best thing to do was to close off the entrance. Then we left."

Fiege spent the next two years trying to obtain Army permission to dig the site. Finally, on August 5, 1961, permission was received from the Secretary of the Army and the U. S. Treasury. Fiege returned to the site accompanied by the commanding general of White Sands, a U. S. Secret Service agent, and fourteen military police.

In front of this audience, Fiege flopped. He couldn't find the cave. He walked up and down the peak and Hembrillo Basin, and he couldn't find it. He took a lie detector test a month later, and it confirmed that he was telling the truth. At that juncture, believing Fiege, the Army set up its full-scale mining operation at the old Noss site. It was illegal, it was secret, and it was lied about. It didn't stop until fiery Ova Noss sent in her agents to get the license numbers.

There is suspicion that the Army was again engaged in surreptitious activities in the 1970s. In 1975, F. Lee Bailey said, "My clients saw some people take two tons of gold from White Sands in two Jeeps and a truck. This was reported to the Assistant Secretary of the Treasury, and no action taken. They are not enforcing the law."

Since the kidnaping and torture of Willie Daught in 1929, the tale of the treasure has provoked dark suspicions. The darkest of these surround the role of the Army.

In 1974, the commanding general of White Sands Missile Range gave an official press briefing to refute F. Lee Bailey's claim of a treasure-trove. The general said:

•There was no gold on the range.
•A 1963 exploration by the Museum of New Mexico had concluded "that no caverns such as claimed by Doc Noss were revealed. . . ."
•"The geologist concluded in his report" that Victorio Peak "contained no traces of gold-bearing ore."

The Army's consistent policy since it seized the Victorio Peak region in the 1940s has been to deny the very possibility of gold in the San Andres. In the course of pursuing this policy, the Army has consistently lied.

For instance, the New Mexico state museum investigation report concluded not that there *weren't* caverns in the peak, but that there *were* and *are* caverns. This conclusion is contained in a letter dated March 12, 1965, from the museum to—yes, the commanding general of White Sands. It states:

"The results of the exploration program conducted in 1963 proved the existence of a number of open cavities within Victoria Peak similar to those described by the individual who claimed to have been in the caves and seen the artifacts and treasure.

"One phase of the program conducted in 1963 consisted of a gravity survey, which indicated a large void or cavity, and a tunnel was started to reach this cavity." However, the digging of this tunnel was stopped by the Army.

The Army's second statement, referring to the geologist's report, is also false. The sole geological report on the peak was done in 1948 by the state Bureau of Mines.[3]

It concluded:

[3] This report was October 13, 1948, and entitled, "Report on Proposed Noss State Mineral Leases, Socorro-Dona Ana Counties, New Mexico."

"The existence of an ore body exposed in Victoria Peak by the caverns along the contact between the limestone and diorite is not only possible but highly probable. . . . Samples taken of the diorite itself where exposed on Victoria Peak assayed a trace of gold. . . . About one-quarter mile east . . . there are prominent outcrops of barite which carry values in copper, lead, and a small amount of silver and gold. . . . To the west of Victoria Peak there is reported to be placer gold in the arroyos adjacent to the dike. . . ."

Whether the Army was purposefully deceptive in its statement will be examined in later chapters. The question to be addressed now concerns the San Andres Range—its geology, its history, and whether gold at any time has been discovered or mined in the region, gold in sufficient amounts to account for the trove.

In pursuit of the answer, we will encounter volcanoes, a Hawaiilike climate, elephant hunters prowling the shores of a recent sea, pre-Columbian priests with electric light machines, conquistadors, Apaches, and the legends of lost mines. Along the way we'll become acquainted with the state of New Mexico—a singular land, an adventurous and beautiful place that attracts and keeps dreamers, and a land that, when viewed through the long end of history's glass, often looks like one vast asylum, its inmates chasing gold.

We will inspect closely that corner of the asylum involving Hembrillo Basin, the Caballos, the San Andres, and the Jornada. We start at the north end of the range, at the crater that shines like glass.

II

The Site

Thunder and lightning rumbled and flashed in the early morning of July 16, when the final test was to be made. The bomb was carefully mounted on a steel tower. Over five miles away, the scientists lay flat, listening breathlessly to the time signals announced over the radio: "Minus 15 minutes . . . minus 14 minutes . . . minus 13 minutes." At "minus 45 seconds" a robot mechanism took over the controls. Suddenly a giant ball of fire rose as though from the bowels of the earth, then a pillar of purple fire, 10,000 feet high, shooting skyward. Its bottom was brown, its center was amber; its top, white. The flash lit up every crevasse and ridge of the San Andres Mountains. Then it shot higher, to 40,000 feet, a huge, rainbow-colored ball, turning swiftly to mushroom shape. It was lit from within by lightninglike flashes. There was a tremendous sustained roar. In Albuquerque, 120 miles away, the sky blazed noonday-bright. When it was done, the tower had completely vaporized. There was only a crater a quarter-mile wide lined with melted rock and sand.

—Eyewitness account of the world's first atomic explosion, at Trinity Site (July 16, 1945)

The region where the gold legends occur is in south-central New Mexico. It is shaped like a parallelogram. The northern boundary is a 69-mile line from Socorro to White Oaks, passing just above Trinity Site. The eastern boundary is the Sacramento Mountains, running 140 miles down to Mexico to meet the Rio Grande River. The Rio Grande, curving eastward, forms the western and southern boundaries.

If you stand on Hembrillo Basin, near the center of the parallelogram, you are in the midst of a time machine. To the west is the Jornada.[1] To the east is the White Sands Desert, formerly the bed of a sea whose shores were grazed by brontosaurus and diplodocus. In this region there appeared the first of the mammal-like dinosaurs, trytyldontoid. Contemporary to them were the first birds, first bony fish, and the ancestors of modern social insects—bees, ants, and wasps.

A remnant of an ancient sea still remains. It is at the foot of the San Andres in the White Sands Desert. It is a "lake" dying but not yet dead. Called Lake Lucero, it is a lingering vestige of the time—only about eight thousand years ago—when elephants, llamalike camels, and full-sized horses roamed the shore simultaneously with Folsom man.

The bed of the original sea is marked on geological maps as Tularosa Basin, an area considerably larger than Delaware, Rhode Island, and the District of Columbia combined. It is a mountain-rimmed valley with no drainage outlet, known to geologists as a "bolson."

The mountains and the sea were formed some ten million years ago. They were not formed in the customary way—by the raising up of the mountains—but by a cave-in of what became the seabed. A severe faulting caused the land between the San Andres and the Sacramentos to collapse, leaving jagged walls on either side, the scarred faces of which became the cliffs of the Sacramento and San Andres mountains.

A huge fresh-water lake formed and sustained itself through successive ice ages until about ten thousand years ago, when a change in the planet's weather—still not completely understood—caused the last retreat of the North American glaciers. The original sea dried up completely about four thousand years ago. It continues to reappear, however, in the diminished form of Lake Lucero whenever there is heavy rainfall in the San Andres.

During rainy periods it is dotted with ponds and pools of water.

[1] This is often mistakenly called Jornado del Muerto. New Mexico historian Fray Angelico Chavez said the correct term is La Jornada del Muerto (The Dead Man's Route). The Jornada was named in 1670 after Bernard Gruber, a German trader who had been imprisoned in Santa Fe on the charges of being a heretic and a necromancer. He escaped the jail but died on the trail as he fled South.

In heavy rains it fills to its brim and beyond. In September 1941, when nine inches of rain fell in a single month, it filled beyond its banks and through much of the Tularosa Basin, giving modern man an impression of the original sea.

The lake water is now undrinkable. It is saturated with gypsum, a white, light mineral closely related to Epsom salt. During the aeons, the gypsum has run off into the lowest areas of Tularosa Basin, where it has created the vast, silent, sugarlike desert of White Sands.

Animal life in the Malpais and on White Sands is not numerous but covers a surprisingly large range of species. At White Sands National Park, service personnel have inventoried 144 species of birds, 23 species of mammals, 371 species of insects, and a few species of reptiles. The mammals include coyotes, badgers, skunks, porcupines, and foxes.[2]

In contrast to White Sands, in the northeast of Tularosa Basin, is a vast black lava bed, the Carrizozo Malpais. It curves down the rim of the basin for forty-four miles. Here, on the malpais, animals have formed special colors. Mice, kangaroo rats, squirrels, prairie dogs, and snakes have developed dark-hued colorations to protect them against the bobcat, the fox, the coyote, the badger, and the occasional black bear that roam the malpais.

The same species of animals that are dark on the malpais have adapted to the glaring whiteness of White Sands by turning themselves light.

The lava field is not ancient. It was formed at the time of Christ, witnessed by the eyes of men when a series of volcanic explosions rocked the valley. The last of the explosions created Little Black Peak, a neatly defined, small volcano with an intact crater. It is currently dormant but, sitting atop deep fissures, it is subject to eruption.

The largest mammals to regularly roam the basin are the coyote and the badger. This odd pair—the coyote lithe and quick, the badger squat but with deadly, powerful claws—have formed an apparently symbiotic relationship.

They are seen traveling together in the desert, the badger

[2] The description of Tularosa Basin comes from Natt N. Dodge, "The Natural History of White Sands National Monument" (Globe, Ariz.: Southwest Parks and Monuments Association, 1971).

shuffling along, eyes intent on the ground, the coyote trotting beside, head up and ears and eyes alert.

Writer J. Frank Dobie reported seeing an ancient bowl from northern Mexico that had a bas-relief of a coyote head on one side and a badger on the other. The bowl had a rounded bottom. When spun, the coyote would chase the badger, and the badger would chase the coyote on a circular trail that had no end.

Both animals are carnivores, living off the prairie dogs, mice, gophers, rabbits, and other burrowing animals in the basin. Apparently they share the spoils of the badger's raid on the burrows. The badger burrows in, the rabbit or whatever bolts through an escape exit, and the coyote runs him down.

The most exotic visitor to the basin, while not a permanent resident, is the coati. This large relative of the raccoon comes up from Mexico periodically, crossing numerous mountain ranges to get to the San Andres. No one knows why.

All of the animals catalogued in the basin roam the San Andres. The climate of the San Andres is arid but allows certain marginal vegetation such as prickly pear, some grasses, buffalo gourds, mesquite, piñon trees, and rarely cactus. You can even find there a plant usually associated with wet, tropical forests. It is a rubber tree. Its local name is "rabbit bush," a shrub with bright yellow blossoms whose sap contains latex rubber of high quality.

The plants attract insects, which attract birds, which attract small ground animals, which attract larger carnivores, whose defecation nourishes plants. The life spins in a circle, one chasing the other like the coyote and the badger on the ancient Mexican bowl.

The one place in the San Andres that the animals avoid is the Hembrillo Basin. And this is a deep mystery. The basin's vegetation seems to be the same as elsewhere; indeed, it is even lusher, for at the base of Victorio Peak is a cool, drinkable fresh-water spring that has not been dry for well over a century. It is one of the few nonalkaline springs in the entire San Andres.

Cowboys in the area say that the Apaches have a legend that the basin is haunted. Strange lights can sometimes be seen there at night. It is these lights—spirits—that keep the animals away.

That is one legend.

The central legend of the San Andres, of Hembrillo Basin, of White Sands, is that they hold gold, more gold than Fort Knox.

Gold is found everywhere on the planet—in the seas, in riverbeds and creeks, in the earth's interior, and throughout the rocks of the earth's crust. It is found in quantities large enough to be profitably mined, however, in only two forms: veins and placers.

Veins occur when gold and other minerals are flushed up as hot liquids from the earth's bowels into the planet's crust. When the minerals encounter cooler rocks near the crust, they crystallize. Usually this occurs within a few thousand feet of the surface. The crystalline material, including gold if present, fills cracks and fractures in the rocky crust and thereby forms veins of gold ore imbedded in rock.

Sometimes, however, the minerals do not encounter cool rock but continue to the surface where, mingled with ground water, they spurt forth onto the planet's surface as hot springs; they cool and settle into nuggets.

Placer gold is created by erosion of the gold veins. The veins are subject, as all rock, to weathering and erosion from wind and water and temperature. Rocks and minerals, including gold, disintegrate into small fragments. These in turn are washed into gullies, creeks, and rivers. Most of the metal-bearing minerals are easily dissolved and become part of the water. Gold, however, is extremely resistant to dissolution and is carried downstream intact as nuggets.

On curves in streams, where the current slows, the gold has time to settle into the gravel and sand. It collects there in pockets and, it is there that it becomes placer gold. It has settled in a place.[3]

Prospectors look for such pockets in both contemporary living streams and in the dry beds of ancient streams. In many areas the discovery of placer gold has been the first clue to gold veins in the mountains at the headwaters.

Doc Noss, and others, reported seeing a stream in Victorio

[3] Placer (pronounced "plasser") has the same root as "plaza" and derives from the plazalike sandbanks where such gold was first found in the American Southwest by the Spanish.

Peak and seeing evidence of the mining of veins. Since then several authorities, most notably the White Sands command, have denied that Victorio Peak contained gold in any form. Of course, motives should be weighed. Noss, if he was a con man, would have a self-serving purpose in claiming the presence of gold when there was none. And the Army, if gold is there and it hopes to protect it, would have purpose in claiming that there was no gold.

An objective, disinterested report of gold possibilities at Victorio Peak came in a 1948 investigation by geologist Donn Clippinger.

At the request of the New Mexico land office, Clippinger had examined Victorio Peak to see if Noss deserved a lease. He said yes, there were good possibilities of a mother lode.

The numerous stories of lost mines in the Southwest fire the imagination and make interesting reading, but generally are nothing on which to base a legitimate mining venture. Disregarding all folklore and fable, the existence of an ore body exposed in Victoria Peak by the caverns along the contact between the limestone and diorite is not only possible but highly probable. . . .

Samples taken of the diorite itself where exposed on Victoria Peak assayed a trace of gold. . . . About one-quarter mile east along the contact of the dike and the limestone there are prominent outcrops of barite which carry values in copper, lead, and a small amount of silver and gold. . . . To the west of Victoria Peak there is reported to be placer gold in the arroyos adjacent to the dike. Time did not permit investigation of this area.[4]

In 1963, the Museum of New Mexico, with federal and state observers present, excavated in an attempt to settle once and for all whether Victorio Peak contained gold—either as a mine or as a cache. In events that will be discussed in later chapters, the Army closed down the project, and the museum was not allowed to complete its work.

Since then the various commanding generals of White Sands have insisted, repeatedly, that there is no gold in the San Andres.

[4] Donn Clippinger (comp.), "Report on Proposed Noss State Mineral Leases, Socorro-Dona Ana Counties" (Socorro, N.M.: New Mexico Bureau of Mines, Oct. 13, 1948).

Furthermore, they say, there is no gold even close to Victorio Peak and never has been.

The Army position is demonstrably false.

It is unknown how much gold was taken out of New Mexico by the Spanish and the Mexicans. Accurate records have been kept on American activity since 1880, however. They show that six of the largest gold mines in New Mexico are, or were, located within seventy-five miles of Hembrillo Basin. These mines are as follows:

•*The Organ Mountains.* This range appears on maps as early as 1682 as Los Organos (Pipe Organs) because of irregular, columnar peaks that resemble organ pipes. The range is 43 miles south of Hembrillo and, ironically, is located *directly* above the headquarters buildings of the White Sands Missile Range. You could stand at the site of one of the old gold mines and with a small mortar lob shells directly into the Army's press briefing room.

The Organ mines were worked in the late nineteenth and early twentieth centuries. They recorded production of 11,435 ounces of pure gold.

•*White Oaks.* This is located in the Sacramento Mountains 75 miles northeast from Victorio Peak. The two sites are separated by the White Sands Desert and the malpais. At White Oaks, rocks of the Cretaceous period are cut by gold-bearing veins. The first mine, and the town, were founded in 1879. Within 25 years, $3 million in gold and silver was recorded being taken from the mines. White Oaks was the setting of a novel on gold-mine towns, Emerson Hough's 1903 novel *Heart's Desire.* White Oaks mines had a total recorded production of 146,500 ounces of pure gold.

•*Nogal.* This is immediately south of White Oaks and 64 miles from Hembrillo Basin. It was an offshoot of the 1879 strike at White Oaks and in the 1880s produced $500,000 in gold ore. Recorded production was 12,580 ounces.

•*Jicarilla.* This is located in a small range of mountains that rise abruptly, on military property, out of the White Sands Desert 44 miles southeast of Victorio Peak. The original mine here was known as "Two-barrel Mine" because the original prospector, S. M. Perkins, traded his entire claim for two barrels of water. (Perkins was captured once by the Apaches, who stripped him naked and prepared to murder him. He was a hunchback, however, and when the Apaches saw his form they freed him because of a superstition that he was blessed.) Despite the loss of Two-barrel Mine, Perkins later discovered several rich lodes. The town of Orogrande sprang up in the Jicarillas when a very large gold nugget was found. By 1914, the town had a population of 2,000, and 8 placer mining companies were active. The mother lode has never been found. Recorded production was 16,500 ounces.

•*Hillsboro*. Located on the west side of the Rio Grande, it is 55 miles due west of Victorio Peak. Discovered in 1877, the Hillsboro district had placers and vein mines. It produced more than $6 million in gold and silver. Recorded pure gold production was 98,000 ounces in placers and 51,000 ounces from ore.

•*Rosedale*. This is located in the San Mateo Mountains, west of the Rio Grande and 68 miles northwest of Victorio Peak. Gold was discovered in the 1880s by a Jack Richardson. He named the town in honor of a lady friend. Recorded production was 27,750 ounces.

Victorio Peak, and Hembrillo Basin, stand approximately at the center of a circle of six of the biggest mining districts in New Mexico. Only four New Mexico districts have produced more gold than Hillsboro or White Oaks. You can start at Victorio Peak and go no farther than 75 miles in six different directions and you find gold or placers in large amounts. Furthermore, geologist Clippinger said there was a high probability of rich gold at Hembrillo itself.

The districts listed above are the officially registered finds, commencing about 1880. But legends abound of more ancient, much vaster treasures in the region. The first, the greatest, and the most enduring of those legends is that of the Seven Cities of Cibola.

III

Cibola

In the year 1530, Nuno de Guzman, President of New Spain, owned an Indian . . . who as a boy had seen towns so large that he could compare them in size to Mexico [city] and its suburbs. There were seven of these towns . . . together they are called the Kingdom of Cibola . . . whole rows of streets inhabited by gold and silver workers. . . . In order to reach these seven towns it was necessary to cross a desert for forty days, where there was no vegetation except short grass about five inches in height, and the direction was to the north.

—Pedro de Castaneda, soldier in the Coronado expedition of 1540

By the year 1536, the Spanish had been on the American mainland for a brief twenty-seven years. In that time, they had seized treasure beyond anything ever seen in the Old World. Cortez had plundered enough Aztec gold to inflame the imagination of Europe. In Peru, the Inca Atahualpa, paying his ransom to Pizarro, had made a pile nine feet high of golden vessels, jewelry, and plates in a room twenty-two by seventeen feet square.

Rumors of even vaster treasures abounded. One of these was Aztec, for although the size of Montezuma's treasury is to this day unknown, legends began in Cortez's time of vast portions of it being lost at sea or hidden in the North. Pizarro had been lured south, as had other conquistadors, by the myth of El Dorado. The story of "The Gilded Man" was supposedly born from a custom of the highly civilized Chibcha Indians of Colombia, who each year nominated a chief and rolled him in gold, which he then washed off in a sacred lake or well. Although the custom had apparently vanished long before the arrival of the Spanish, its tale had grown into a legend of a land of gold and plenty.

Another myth, that of the Seven Cities of Cibola, drew men to the North. Here the streets would be paved with gold and the doors of high houses studded with precious gems.

Strangely, the myth of the Seven Cities of Cibola was prevalent in Spain four hundred years before Columbus' voyages. The story stated that in the eighth century seven Spanish bishops, fleeing the approach of invading Moors, fled from Oporto westward across the ocean, where they founded seven cities. One Portuguese sailor asserted that he had seen this paradise, where even the sands were one-third pure gold.

When the Spanish arrived in Mexico, they found an ancient Indian legend of "seven caves" to the north, where the Aztecs and other tribes had their origin. This myth was reinforced, according to an account written in 1584 by Balthazar Obregon, when Cortez discovered among the possessions of Montezuma some chronicles, drawings, and paintings that correctly revealed the origin of the Aztecs to be in the North.

These two myths dovetailed and complemented one another. By 1536 the cities of Cibola, "seven underground cities" or caves, had become the gossip of the New World. Cibola was the symbol of gold, wealth, and power, and, as with many myths, there was a basis of reality. The Seven Cities of Cibola existed.

Cibola is the oldest place name in New Mexico. It appears on Castillo's map of 1541 as "la Ciudad de Cibola" and is shown with towers and walls on a somewhat smaller scale than that of "la Ciudad de Mexico," the Aztec capital. According to the two foremost historians in the field, the name was learned by the Spanish from the Pima Indians and may be a transliteration of Shiwina or Shi-uo-na, the Zuñi name for their tribal hunting ground in western New Mexico.[1] There the Zuñis had, yes, seven stone cities of several thousand in population. The ruins still stand, and some of the cities are still inhabited. The first written

[1] The authorities referred to are Frederick Webb Hodge and Adolph Bandelier.

mention of the Zuñi Cibola came in 1539 by Friar Marcos de
Nizza. He called it "Ceuola," very close to Shi-uo-na.

In the year 1536, Nuno de Guzman, governor of New Galicia,
was standing on the balcony of his house in the tiny capital of
Compostela in central Mexico near the Pacific Coast.

He was watching for the arrival of four men being brought to
him under escort. These four were Cabeza de Vaca, Andres
Dorantes, Alonso Maldonado, and Estavanico, Dorantes' black
Moorish slave.

They were the first Europeans to cross North America from
Florida to the Gulf of California. Guzman waited to interview
them on what they had seen. He had already received advance
tales that the four had encountered legends of a nation called
Cibola, which was richer than the Aztecs.

Cabeza de Vaca had come to the New World from Spain in
1527. In 1528 he participated in an expedition to Florida under
Panfilo Narvaez. Landing in Tampa Bay, the expedition of six
hundred men was soon scattered by hostile and powerful Indian
tribes. The Spaniards fled to the coast, built crude barges, and
took to the sea.

De Vaca and the other three eventually shipwrecked somewhere
in the six-hundred-mile stretch of coastline between Mobile Bay
and Galveston.

Their adventures during the next eight years, written by de
Vaca, the leader of the group, form one of the most remarkable
stories in the annals of exploration. The men wandered always
westward, sometimes together, sometimes alone. The tribes encountered by de Vaca and his companions varied widely in custom and languages, but most of them had a common denominator: They were miserably poor.

The four men were at various times held as slaves, beaten with
sticks, and made to do the meanest work. Some Indians amused
themselves by plucking out the hair of the Spaniards' beards. But
misfortune and fortune fluctuated. Mistreated as slaves in one
tribe, they were made friends in the next. Mostly they starved, like
the Indians, living on roots, worms, spiders, and lizards. Several
times they fell seriously ill. There was, however, every year the
"time of the full bellies." This was when the prickly pear was ripe.

In 1973 Ova Noss showed these objects to Albuquerque newsmen. She said they had been taken from Victorio Peak by her husband, Doc Noss, in the 1930s. They include some spillover gold, presumably from a smelting operation; a sword; a sheath; and a cup (photographs 1, 2).

These photos were copied from Ova Noss's family album. The upper shows Doc Noss standing outside the cave entrance at Victorio Peak. The lower is a formal portrait of Doc Noss (photographs 3, 4).

Also taken from Victorio Peak by Doc Noss. According to his widow Ova, these were ingots, one marked LA RUE, all of which have since disappeared; a silver napkin ring inscribed "Talbot Hall, November 17, 1868," and a two-handled cup with the word "Brazil" inscribed on it. The ring and cup are still in Mrs. Noss's possession (photographs 5, 6, 7).

March 1949: Doc Noss lies dead in this photo taken thirty minutes after he was shot by his partner, Charlie Ryan, in a dispute over another attempt to remove the gold from Victorio Peak (photograph 8).

The sweet-tasting and highly nutritious plant was eaten raw, boiled into jelly, and dried into candy.

As the years passed, the men gradually won their freedom. De Vaca earned his first, hitting on the device of being a merchant:

"My stock consisted mainly of pieces of seashells and cockles and shells with which they cut a fruit which is like a bean, used by them for healing and in their dances and feasts. This is of the greatest value among them, besides shell beads and other objects. These things I carried inland, and in exchange brought back hides and red ochre with which they rub and dye their faces and hair; flint for arrow points, glue and hard canes wherewith to make them, and tassels made of the hair of deer, which they dye red. This trade suited me well. I was not bound to do anything and was no longer a slave!"[2]

The Indians were not precisely noble savages. Of the inhabitants of an island where de Vaca spent six years, he wrote:

"There were two tribes on the island, the Capoques and the Han. They have the custom when they know each other and meet from time to time, before they speak, to weep for half an hour. After they have wept, the one who receives the visit rises and gives to the other all he has. The other takes it, and in a little while goes away with everything."

Ethnologists to this day are searching for the identity of the Capoque and the Han. Very few of the tribes encountered by de Vaca have been identified. One problem is that he wrote the names phonetically, using Spanish sounds for Indian words, and it is unknown if the names he used were the tribes' names for themselves or the names given them by other tribes.

Whatever their names, they were a miserable people. Of an inland tribe called the Iguaces, de Vaca wrote:

"Their principal food are two or three kinds of roots, which they hunt for all over the land; they are unhealthy, inflating, and it takes two days to roast them. Many are very bitter, and with all that they are gathered with difficulty. But those people are so much exposed to starvation that these roots are to them indispensable and they walk two and three leagues to obtain them. Now and then they kill deer and at times get a fish, but this is so

[2] Fanny Bandelier (trans.), *The Journey of Alvar Nunez Cabeza de Vaca* (New York: Allerton Book Company, 1904).

little and their hunger so great that they eat spiders and ant eggs (the pupas), worms, lizards, and salamanders and serpents, also vipers, the bite of which is deadly. They swallow earth and wood, and all they can get, the dung of deer and more things I do not mention; and I verily believe, from what I saw, that if there were any stones in the country they would eat them also."

De Vaca and his party were the first white men to see the buffalo, which he called "cows":

"I have seen them thrice and have eaten their meat. They appear to me of the size of those in Spain. Their horns are small, like those of the Moorish cattle; the hair is very long, like fine wool and like a peajacket; some are brownish and others black, and to my taste they have better and more meat than those from here [meaning Mexico]. Of the small hides the Indians make blankets to cover themselves with, and of the taller ones they make shoes and targets [that is, shields]. These cows come from the north, across the country farther on, to the coast of Florida, and are found all over the land for over four hundred leagues."

During one of his enslavement periods, de Vaca became a tanner:

"They would make me scrape skins and tan them, and the greatest luxury I enjoyed was on the day they would give me a skin to scrape, because I scraped it very deep in order to eat the parings, which would last me two or three days. It also happened to us, while being with these Indians and those before mentioned, that we would eat a piece of meat which they gave us, raw, because if we broiled it the first Indian coming along would snatch and eat it. . . ."

Children were nursed at the breast until the age of twelve. The practice was common among many of the tribes, and de Vaca was told that it was due to the perennial scarcity of food. "It was common (as we saw) to be without food two or three days and even four, and for that reason they nursed the little ones so long to preserve them from perishing through hunger."

Another practice common among the Indians was an intoxicating smoke. "In this whole country," wrote de Vaca, "they make themselves drunk by a certain smoke for which they give all they have." It is unlikely that he is referring to tobacco. The leaf to that plant had already been introduced to Spain, and de Vaca was probably familiar with it. He also states that the Indians got "drunk," and de Vaca was a precise reporter. Tobacco doesn't intoxicate. There are forms of hemp indigenous to North America, and it may have been one of these. Hemp is the plant from which marijuana and hashish are obtained.

Homosexuality was also present, to de Vaca's disgust: "During the time I was among them I saw something very repulsive, namely a man married to another. Such are impotent and womanish beings, who dress like women and perform the office of women, but use the bow and carry big loads. Among these Indians we saw many of them; they are more robust than the other men, taller, and can bear heavy burdens."

Later historians would credit de Vaca with triggering the rush to Cibola by his repeating stories of cities to the north where could be found gold and silver. It is a widespread and curious mistake, because de Vaca made no such claim in either of his two writings on the journey, the *Relacion* and the longer *Los Naufragios*. Indeed, the viceroy of New Spain, Don Antonio de Mendoza, in a friendly letter to the King regarding de Vaca, does not seem to attach much practical importance to the adventure.

De Vaca's first mention of a valuable object encountered was a copper rattle given to Dorantes. "A big rattle of copper, large, on which was represented a face, and which they held in great esteem. They said it had been obtained from some of their neighbors. Upon asking them whence it had come, they claimed to have brought it from the North, where there was much of it and highly prized. We understood that wherever it might have come from, there must be foundries, and the metal was cast in molds."

By the time de Vaca's party reached the Rio Grande, they had acquired awesome reputations as medicine men, and their journey from that point onward became a triumphal procession. They were called "Children of the Sky" and "Children of the Sun."

Thousands of Indians joined their train, and each tribe they encountered pressed upon them gifts of food, clothing, and other objects.

At the Rio Grande they encountered their first pueblos, stone houses. "These were the first abodes we saw that were like real houses." Here, and for the remainder of the journey, the Indians were wealthy; food was plentiful, and agriculture was thriving, particularly of maize, but also of beans and squash. All of the Indians encountered prior to the Rio Grande had usually been naked, even in the coldest winters. At the pueblos, however, the people wore cotton clothes and softly tanned deerskin. They wore skirts and shirts and shoes.

Far from saying that the pueblos were cities of gold, de Vaca reported:

"The Indians who live in permanent houses [that is, Pueblos] and those in the rear of them pay no attention to gold nor silver, nor have they any use for either of these metals."

The Spaniards traversed the Pueblo area and then turned south, toward Mexico City. At a site that is known to be Corazones, in Mexico, they saw emeralds.

By coincidence, it was Dorantes who again was the recipient.

"Dorantes was presented with five emeralds, shaped as arrow points, which arrows they use in their feasts and dances. As they appeared to be of very good quality, I asked whence they got them from, and they said it was from some very high mountains toward the north, where they traded for them with feather bushes and parrot plumes, and they also said that there were villages with many people and very big houses."[3]

After passing the pueblos, the party came to a river where a tribe told them of other white men—bad white men. In March 1536, on the Rio de Potatlan in Sinaloa, they walked into the camp of Captain Diego de Alcaraz, a cruel and vicious man on expedition to illegally gather slaves. (Alcaraz tried to enslave an honor guard of six hundred Indians accompanying de Vaca, but de Vaca forced him to desist.)

De Vaca wanted to go to Mexico City to tell his story, and his journey would take him through Compostela to be interviewed by Governor Guzman. Along the way, Alcaraz questioned de Vaca about gold in the North. De Vaca denied it, but Alcaraz would

[3] Most writers on de Vaca believe the "emeralds" were some other stone, most likely garnets or peridots. But in a report of 1893, Governor Thornton of New Mexico noted that there was then in Santa Fe a cut emerald of fine quality weighing one carat and that "many other smaller ones have been found." F. A. Jones in 1915 reported that a few emeralds of good quality had been found in gravels near Santa Fe. Therefore, though rare, emeralds do exist in at least the Santa Fe region. See Stuart A. Northrop, *Minerals of New Mexico* (Albuquerque: University of New Mexico Press, 1959).

later accuse de Vaca of keeping his real knowledge secret so that he might return on an expedition of his own.

As Guzman waited for de Vaca he recalled a story told him years earlier by Tejo, an Indian who was Guzman's servant when Guzman was governor of Panuco.

Tejo, on the basis of his name, may have been a Teja Indian, a tribe discovered by Coronado and from which Texas takes its name.

Pedro de Castaneda, Guzman's aide, said Tejo "was the son of a trader deceased long ago, who, while his son was still a child, was wont to travel over the interior of the country in order to sell the handsome plumes that are used for headdresses by the Indians. In exchange he brought home a large quantity of gold and silver, both metals being very common in that region. He added that once or twice he had himself accompanied his father, and had seen towns so large that he could compare them in size to Mexico [City] and its suburbs. There were seven of these towns, and there were whole rows of streets inhabited by gold and silver workers. He said besides that in order to reach these seven towns it was necessary to cross a desert for forty days, where there was no vegetation except short grass less than half a hand-span in height and the direction was to the north between both oceans [that is, the gulfs of Mexico and California]."[4]

Finally de Vaca and his party approached. Guzman had them as his guests that night and was told of the large pueblos to the north, but told of no gold. Like Alcaraz, Guzman believed de Vaca to be hiding something. He reflected on Tejo's story and decided to mount an expedition to the Seven Cities of Cibola. For political reasons, he could not lead it himself, and Tejo could not guide it. Tejo had died five years earlier.

The man chosen to lead the expedition was Francisco Coronado. He would be accompanied by a priest, again for political reasons but also because the soldiers needed a priest, and priests frequently got along well with Indians.

The priest was Fray Marcos of Nizza, probably an Italian from Nice. He had acquired experience and prestige while participating

[4] A. Grove Day, *Coronado's Quest* (Berkeley: University of California Press, 1964).

in Pizarro's conquest of Peru. There the friar had witnessed the murder of the Inca Atahualpa and presumably held aloft the cross on that Christian occasion.

As his guide, Marcos chose the Moorish slave Esteban. Coronado led the expedition no farther than Corazones, where he ran out of food and had to stop. Fray Marcos went forward with a smaller body.

Esteban was delighted to be in the company. In the later stages of his journey with de Vaca, Esteban had been in his glory. Their march across the Southeast had been a triumph. At each village, the inhabitants had presented them with all their possessions as gifts. Esteban was the one to march in advance, leading the Children of the Sun to the land of the sun. "He it was who organized the reception at each new town on the route, gleaning information about the villages ahead, speaking with a drawl the dialects he had picked up, grinning at the shy Indian womenfolk. He had pranced ahead always, an ebon major-domo, beating a drum of hide and shaking a bright-feathered gourd rattle, gift of a medicine man of the plains. Esteban still had that rattle; he knew its power over the Indians. He had been himself a noted medicine man, a maker of magic."[5]

Fray Marcos chose Esteban to lead an advance party and spread word to the Indians of the coming of the Spanish. He was a black man out front alone with a few Indians but no Spaniards. They would follow by several miles, eventually by several days. Marcos told Esteban "to go to the North fifty or sixty leagues [approximately 150 miles] to see if in that direction there might be observed something great, or some rich country and well settled; and if he found anything or heard anything of that kind to stop and to send me a message by some Indians. That message was to consist of a wooden cross of a white color. In case the discovery was of *medium* importance, he was to send me a cross of one span in length; and if *more important than New Spain,* he should send me a large cross."[6]

Esteban proceeded adorned—it was his own idea—with rib-

[5] Ibid., p. 19.

[6] Adolph Bandelier, "Fray Marcos of Nizza," Papers of the Archaeological Institute of America, American Series V (Cambridge, Mass.: 1890).

bons and cockades, bright feathers in his hair, and draped with all kinds of whistles, rattles, bells, and drums. His costumery had great impact upon the Indians, particularly women. Many left their tribes to join Esteban, dropping out from their cultures to follow a pied piper.

Almost from the beginning, crosses were being sent back to Fray Marcos. They were so huge that their weight bent the backs of two Indians. Both parties pressed on at great speed as Esteban received intelligence of gold and great cities from the Indians. And throughout the Spanish exploration period, the Indians told their listeners precisely what they wanted to hear. There is gold? Yes, there is gold. There is no gold? No, there is no gold.

Meanwhile, after forty days, Esteban actually had reached the city identified by the Indians as Ahacus. It was probably the Zuñi pueblo of Hawikuh, now in ruins 120 miles due west of Albuquerque.

Esteban's first messenger to Marcos told the friar, "At the foot of the mountains there is a large and mighty plain, wherein they told me, there were many great towns and people clad in cotton: and when I showed them certain metals which I carried with me, to learn what rich metals were in the land, they took the mineral of gold and told me that thereof were vessels among the people of that plain, and that they carried certain round green stones hanging at their nostrils . . . and they have certain thin plates of that gold, wherewith they scrape off their sweat, and that the walls of their temples are covered therewith, and that they use it in all their household vessels."

Having been a participant in the Pizarro expedition, Marcos had experienced vast amounts of gold and, accordingly, the Indian's tale was not so improbable as it might seem to others.

On the following day, another Indian messenger came with a similar report and added that Esteban was coming near to "a province called Ceuola . . . with seven great cities in this Province, all under one Lord, the houses whereof are made of lime and stone, and are very great, and the least of them with one loft above head, and some two and of three lofts, and the house of the Lord of the Province of four, and that all of them join one onto the other in good order, and that in the gates of the principal houses there are many Turquoise stones cunningly wrought." This was all true, of course, even to the studding of house entrances with turquoise.

Marcos, however, was a bit skeptical. "I told them it was impossible that the houses should be made in such sort as they informed me, and they for my better understanding took earth or ashes and poured water thereupon and showed me how they laid stones upon it, and how the building grew up, as they continued laying stones thereon, until it mounted aloft. I asked them whether the men of that country had wings to mount up into those lofts; whereat they laughed and showed me a ladder."[7]

As he approached closer to Cibola, he encountered a townsman of that province. He was "a white man of a good complexion: and a man of prestige among the local Indians. He said that the houses are of lime and stone, according as other had told me before, and that the gates and small pillars of the principal houses are of turquoises and all the vessels wherein they are served, and the other ornaments of their houses were of gold, and that the other six cities are built like onto this, whereas some are bigger, and that Ahacus is the chiefest of them."

There was no welcome. Instead, Esteban was taken prisoner, and all his bells and objects for exchange were seized with him. The next morning he and some three hundred of his followers were surrounded by a cordon of Zuñi who bludgeoned the party to death. One Indian escaped, and Marcos set down his words:

"He told me that one day previous to reaching Cibola, Esteban sent, as he was wont to do always, his gourd, in order to show them in what quality he was coming. The gourd had a few strings of rattles and two plumes, one of which was white and the other red. When they reached Cibola and presented the gourd to the person whom the lord has placed there in charge, he took it into his hand, and seeing the rattle with great wrath threw the gourd on the floor, and said to the messengers that they should forthwith leave the town, that he knew what kind of people those (the strangers) were, and that they should tell them not to enter the place lest they should all be killed. The messengers returned and reported to Esteban what had happened, who said that this was

7 Ibid., pp. 203–31.

nothing—that those who at first displayed anger always received him in the kindest manner.

"So he continued his road until he reached the city of Cibola where he met people who refused to allow him to enter, and placed him in a large house outside, taking from him all he carried of objects for exchange, turquoises, and other things received from the Indians on the journey. There he was all night, neither food nor drink being given to him nor to his escort.

"On the following morning, this Indian (the one telling the tale) felt thirsty, and went out of the house to get a drink of water at a stream near by, and a short while afterwards he saw Esteban endeavoring to escape, pursued by the people of the city, who were killing some of the people of his company. Seeing this, this Indian concealed himself and crept off stealthily up the said stream, and finally crossed over to take the road through the desert."[8]

After hearing of the massacre, Marcos—by his own account—continued on. He came, he said, to a hill from which he had sight of the first, and smallest, city of Cibola. It had the name of Ahacus and was the place where Esteban was killed.

Wrote Marcos, "I followed my road till we came in sight of Ceuola, which lies in a plain on the top of a roundish hill. It appears to be a very beautiful city, the best that I have seen in these parts. The houses are, as the Indians had told me, all of stone, with their stories and flat roofs. As far as I could see from a height where I placed myself to observe, the settlement is larger than the city of Mexico. At times I was tempted to go to it, because I knew that I risked nothing but my life, which I had offered to God the day I commenced the journey; finally I feared to do so, considering my danger and that if I died, I would not be able to give an account of this country which seems to me to be the greatest and best of the discoveries.

"As far as I could discern . . . the people are somewhat white, they wear apparel and lie in beds, their weapons are bows, they have emeralds and other jewels, although they esteem none so much as turquoises. . . . They use vessels of gold and silver, for they have no other metal, whereof there is greater use and more abundance than in Peru, and they buy gold for turquoises in the

[8] Bandelier, op. cit.

province of the Pintados, where there are said to be mines of great abundance. . . ."[9]

With the aid of the Indians, Marcos made a heap of stones and placed a small, slender cross on top. Then, in the names of viceroy and Emperor, he laid claim to all the seven cities and the kingdoms of Totonteac and Acus and Marata. He christened the land the New Kingdom of St. Francis.[10]

Fray Marcos returned to Mexico City, where his report on Cibola excited great interest: "More gold than Peru!"

An army of conquest was formed under the leadership of the thirty-year-old conquistador Coronado, who had replaced Guzman as governor of New Galicia.

Coronado set out in 1540 from Compostela, crossed modern Sonora and southeastern Arizona, and entered Cibola itself. He studied it from the heights for a day; then he attacked.

A superior force of Indians awaited him outside the city. But once again that mysterious phenomenon of the conquistador showed itself. The small body of Spanish prevailed, driving thousands of Indians before them, killing hundreds, and losing less than half a dozen men themselves.

Coronado then assaulted the city, himself taking the lead. A conspicuous figure in gold-painted armor, he was twice knocked to the ground by stones, and an arrow struck into his foot. The city of Ahacus surrendered.

There was no gold.

One of the Spaniards recorded, "There we found something we prized more than gold or silver, namely, much maize, beans, and chickens larger than those here of New Spain, and salt better and whiter than I have ever seen in my whole life."

[9] Ibid., Day, op. cit., pp. 53–54.
[10] It was later learned by Coronado that these were actual places. Ahacus was the pueblo of Hawikuh. Marata, to the southeast of the Zuñi nation of "Cibola," was a group of towns that had been recently destroyed at the time of Esteban's visit. Totonteac was an abandoned town near the Hopi towns of the Painted Desert in Arizona. Acus was the citadel rock of Acoma in central New Mexico. Acoma is the most spectacular of the lot, sitting atop a mesa in the middle of a cliff-lined valley. It is the Gibraltar of the desert, and in many places the sides of the mesa are not merely perpendicular but actually overhang. The village is on the summit, approximately seventy acres in extent, and little changed from Coronado's time. It is the oldest continuously inhabited town in the United States.

Coronado sent out lieutenants in all directions. Two of these discovered the Grand Canyon. A third explored the Rio Grande southward to the area of Mesilla and turned back. He very possibly used the *Jornada del Muerto* on the return.

Fray Marcos was branded a liar. There was no gold. And the angry soldiers doubted if he had been within 150 miles of Cibola. Marcos would die disgraced.[11]

A discovery in the desert aroused excitement. One of the men had found a monstrous horn, six feet long and as wide as a man's thigh at the base. Some thought it was an elephant tusk. It was—that of the long-extinct mammoth.

They also found some jewels—turquoises of sky-blue tinge. These were held sacred by the Pueblo tribes and, when polished and decorated, were the chief articles of decoration. Larger pieces were embedded in lintels beside house entrances.

The Indians were careful not to let the Spanish discover the site of the turquoise mines, which were near present Santa Fe. There, at Mount Chalchihuitl, immense pits had been cut in the solid rock of the mountain. Tens of thousands of tons of ore had been broken out and moved barehanded with only stone tools. In the largest pit, a number of tunnels had been run into the rock heart where there were numerous veins of turquoise, as well as strips of gold-bearing quartz. The mountain was honeycombed with these tunnels.

Near the mountain was Pecos, the largest of all the pueblos at that time. There, too, gold was rumored to be in great quantities. Coronado went after it and found a town of more than two thousand people atop a mountain overlooking a beautiful rivered valley. There were two great pyramids of towns, each containing more than five hundred rooms and perfectly fitted for defense.

They had heard of the Spaniards' ferocity, however, and the people of Pecos welcomed them and gave them presents of food and turquoise. But there was no gold.

There was, however, a fascinating generator to make light,

[11] His later years were extremely impoverished. He begged an old friend, the bishop, to give him a little monthly donation of wine "of which I am in great need, as my sickness is of lack of blood and natural heat." The bishop responded nicely, ordering that each month an *arroba* of wine was to be delivered to the old explorer by Indian messenger. Marcos died in 1558 in Mexico City, twenty years after he had left that spot with such great hopes.

which the Indians at Pecos kept secret from Coronado. This was luminescent quartz. It consisted of a rectangular base of pure white-vein quartz with a groove in it and a bolster-shaped upper piece of the same material. Rapid friction by rubbing produced a strong glow in the dark, which was used to light the sacred kivas. The machine still worked perfectly when it was discovered by archaeologist Alfred Kidder in the Pecos ruins, as he reported in 1932. Archaeologist S. H. Ball remarked upon it, "Here we have a perfected machine perhaps seven hundred years old; the first Indian to observe the luminescence of quartz must have done so centuries earlier." (Similar "lightning machines" or "glow stones" have been found at several other localities in north-central New Mexico).[12]

Coronado would eventually journey to the center of Kansas in search of gold, but never find it. The Spanish, however, had found New Mexico and a civilization of Indians.

The Spanish would return in 1598 to colonize the land, making Santa Fe their capital. It was peaceful for eighty-two years, and during that period gold mines were found and dug. Indians were conscripted to do the labor and, partly in reaction to that, in 1680 the Pueblos staged a rebellion of such ferocity and fire that it is almost without parallel in North America. Under the leadership of an extraordinary medicine man named Popé, the tribes united and took up arms. They slaughtered the Spanish, overrunning everything, killing more than four hundred in Santa Fe alone. They chased nearly three thousand of the hated conquerors south to El Paso, cleansing New Mexico entirely of Spanish.

Popé began his own despotic rule and commenced a campaign to wipe out all Spanish traces. Santa Fe was converted to a pueblo. The churches were destroyed, the Spanish language forbidden, and the mines were filled and hidden. Baptized Indians were even washed to rid them of the baptism.

Internal dissension and Apache raids soon weakened Pueblo unity and the Spanish returned in 1692, shortly after Popé's death. Part of the agreement they made with the Pueblos was that no Indians would be used in mine labor, and no mines would be opened.

[12] Northrop, op. cit., p. 8.

In the meantime, the legend of Cibola lived on. For nearly three centuries, prospectors would scour the mountains and deserts of New Mexico looking for the legendary cities of gold. Indeed, in the 1970s, the claimants at White Sands developed a theory that the seven cities of gold actually referred to seven gold mines in the Organs, the Caballos, and the San Andres.

IV

Lost Mines

My clients claim some of the coins are stamped "La Rue." But it represents only a small part of the total gold. What we have there is one hundred tons of gold. . . . Give me just thirty minutes and a helicopter and I can lead the commanding general right to the treasure and show it to him.

—Interview with lawyer F. Lee Bailey (1975)

"La Rue" refers to an obscure legend concerning a French priest who led fugitive Indians into the White Sands-Mesilla area in the early 1800s and found the world's richest mine.

Primary-source documentation of La Rue's existence is lacking. And two of New Mexico's leading authorities are convinced he never did exist. They are state historian Myra Ellen Jenkins and Church historian Fray Angelico Chavez, both of Santa Fe.

Fray Chavez, who inspected likely Church documents, said he could find no mention, direct or indirect, of a Padre La Rue in that period. Dr. Jenkins is equally firm. She says, furthermore: "there is no way for gold to be at White Sands unless it was brought there by F. Lee Bailey and his gang from somewhere else."[1]

The La Rue legend nevertheless has persisted in written form for more than half a century. Its oral tradition is much older and is particularly strong among Indian and Mexican inhabitants of Dona Ana County, the Mesilla Valley-White Sands region.

Legends have two interesting characteristics: One is that they

[1] Interview with author (Aug. 1976).

grow in grandeur the farther they are removed in time and space from their source of origin; the other is that they almost always have *some* substance in truth. Even the tales of Cibola with its "houses studded with emeralds" and "its streets of jewel workers" had a basis of reality: lintels studded with turquoise and street artisans working turquoise.

Legends of gold in the White Sands region are numerous and date back to the early nineteenth century. They vary widely in the sources of the treasure, but all state that there is, or was, a phenomenal mine and a phenomenal cache in the area.

In the main, the legends deal with the Spanish-Mexican period of New Mexico history, a period in which historians disagree sharply on whether or not gold was mined.

The most prestigious voice on the subject is that of the great ethnologist Adolph Bandelier, who advised caution on early Spanish mining stories.

Bandelier is to New Mexico history what Herodotus was to Greek history. He is the beginning and premier ethnologist, who gathered information by travel and personal experience. In the 1880s and 1890s, often accompanied by photographer and friend Charles Lummis, Bandelier walked an estimated ten thousand miles to investigate ruins.

Lummis recalled, "We had no endowment, no vehicles. Bandelier was once loaned a horse, and after riding two miles, led it the rest of the thirty. So we went always by foot; my big camera and glass plates in the knapsack on my back, the heavy tripod under my arm; his aneroid, surveying instruments, and satchel of the almost microscopic notes which he kept fully and precisely every night by the campfire (even when I had to crouch over him and the precious paper with my waterproof focusing cloth) somehow bestowed upon him. Up and down pathless cliffs, through tangled canyons, fjording icy streams and ankle-deep sands, we travailed; no blankets, overcoats, or other shelter; and the only commissary a few cakes of sweet chocolate and a small sack of parched popcorn meal. Our lodging was the cold ground."

Bandelier states flatly that "In New Mexico there were no mines (worked by the Spanish) until after 1725, and compulsory labor on the part of the Indians even after that date was limited to service in the missions."

But there are contradictory statements. The conquistador Chamuscado reported in 1581 finding "eleven silver mines of very rich veins, the ore of three of which was brought to this city [Mexico] and given to His Excellency . . . the assayer found one sample to run 50 per cent silver. . . ."[2]

And Don Pedro Pino, from New Mexico, in reporting to a Spanish congress at Cadiz in 1812, said: "In this province mines have been found closed, some of them with work tools inside; but it is not known at what time they were discovered and worked. There are many mineral veins in the mountains of gold and silver."[3]

And, finally, one of the largest mines on the continent is positively known to date from Spanish times. It is, however, not a gold mine but a copper mine, the Santa Rita in Grant County, which in 1807 Pike recorded as yielding twenty thousand muleloads of metal annually. "The mine," said Pike, "was known in early Spanish times and continued digging has made it one of the greatest open-pit excavations in the world."[4]

The principal cause of the dispute over mining and no-mining derives from the destruction of records in the Pueblo revolt of 1680. Professor John Graham of the Texas College of Mines wrote that "work in the mines, a form of work foreign to the Pueblo experience, was the principal cause of the revolt, and when the Spaniards were allowed to return it was expressly stipulated that mining should be prohibited. The records of such mining as had been done were destroyed in the uprising and the mines themselves, probably only open cuts rather than underground workings, were filled in and all signs of their existence obliterated."[5]

In his definitive work *Minerals of New Mexico,* Professor Stuart A. Northrop reports considerable evidence of Spanish-era mining. The earliest report he found of an actual gold sighting is in 1582 by Friar Augustin Rodriguez, who saw "very good veins, rich in contents. . . . There are so many deposits that it is indeed marvelous."

[2] Quoted in R. E. Twitchell, *Leading Facts of New Mexico History,* Vol. I (Santa Fe), p. 258.

[3] Benjamin Read, *Illustrated History of New Mexico* (Santa Fe, 1912).

[4] T. M. Pearce, *New Mexico Place Names* (Albuquerque: University of New Mexico Press, 1965), p. 149.

[5] John Graham, "Mines and Miners" (El Paso *Herald,* May 25, 1936).

In the period 1620–26, rich silver ore from Santa Fe was being assayed in Mexico City.

In 1630, referring to the area of the "Socorro Mines" presumably adjacent to or in the San Andres, Friar Alonso de Benavides in his *Memorial* notes the existence of "very great treasures of mines, very rich and prosperous in silver and gold," especially in the Socorro region. He says he made numerous assays.

Without comment or judgment, Northrop also reports the following traditions:

Prior to 1680, the Spaniards had mines in Taos country for gold, silver, and copper, using Indian labor. It was rumored that "the Franciscan Fathers, working in the name of the church, are said to have taken out several million dollars in this way."

And in another report, " . . . millions in gold and silver were taken out of the mines in the Taos mountains near Arroyo Hondo, and the Spaniards concealed a hoard of $14 million in one shaft before fleeing [the revolt]."

Northrop recounts many such reports in the Pueblo country.

In the region surrounding Hembrillo Basin, he notes that there are several references, chiefly from articles in the Las Vegas *Mining World* (1880–84), to old Spanish workings at Socorro Peak. Several ancient smelters and mines were found. The slag yielded "both gold and silver."

In Sierra County, it was reported that the Spanish found a rich mine in 1655. "Many rich shipments of gold were made from this mine until 1712."

Jumping to more modern times, in 1846, Dr. F. A. Wisliznus conducted a military reconnaissance of New Mexico for the U. S. Army. He reported much evidence of Spanish mines:

"Great many now deserted mining places in New Mexico prove that mining was pursued with greater zeal in the old Spanish times than at present. . . . The mountainous parts of New Mexico are very rich in gold, copper, iron, and some silver. . . ."[6]

Finally in 1883, old workings and ruins of smelters used by the Spanish were found in the Organ Mountains at the site of the Modoc mine. The Modoc mine is directly above the present headquarters of White Sands Missile Range.

[6] Stuart A. Northrop, *Minerals of New Mexico* (Albuquerque: University of New Mexico Press, 1959), pp. 13–27.

The statements of eyewitnesses to heavy mining activity in the seventeenth, eighteenth, and nineteenth centuries carries convincing weight that the modern historians are wrong when they say "no mining" in the pre-American period.

Therefore, mining having occurred, there is a basis for the gold legends.

For a survey of those legends we turn first to J. Frank Dobie, a Texas native, historian, and folklorist who taught in a number of universities and was a professor at Cambridge University. He gathered oral traditions of lost treasures throughout the Southwest and printed many of them in his book *Coronado's Children*.

Dobie's interviews ranged from Louisiana, where he found tales of treasure buried by the pirate Lafitte, to Southern California, where he encountered myths of fabulous lost mines.

The survey here is restricted to his interviews concerning myths within 150 miles of Hembrillo Basin.

A tradition of gold in the Guadalupe Mountains, to the southeast of White Sands, dates to the nineteenth century. One of the early tellers of it was General Lew Wallace, lawyer, novelist, military man, politician, and romantic historian. While governor of New Mexico Territory (1878–81) he claimed in a written article to have dug out of the basement of the Palace at Santa Fe an ancient document.[7]

The document, according to Wallace, recited how in the early seventeenth century a Christianized Indian had conducted a certain Captain de Gavilan and thirty other Spaniards to a rich gold deposit in mountains to the southeast, which Wallace presumed to be the Guadalupes. The Spaniards mined the site, and on the basis of volcanic evidence called it Sierra de Cenizas (Ashes Mountains). They took out masses of nuggets and very rich ore.

Then came the uprising of 1680, in which the Indians killed all the Spanish who did not flee New Mexico. Sierra de Cenizas has been lost ever since—lost on maps, lost in geography, lost as a mine.

But maybe not.

Dobie quotes the Apache leader Geronimo as saying that the

[7] The Palace of the Governors, standing at the center of Santa Fe's town square and now a museum, was built in 1610 and has been repeatedly restored. Wallace wrote his best-selling classic *Ben Hur* in one of its rooms.

richest gold mine in the world lay hidden in the Guadalupes, a day's horseback ride south from the Mescalero Apache reservation. Dobie also described how a penniless prospector, William Colum Sublett, came into an Odessa, Texas, saloon, ordered drinks for the house, and threw onto the bar a buckskin pouch full of gold nuggets.

For years thereafter, Sublett sneaked into the Guadalupes every few months, eluding trackers. Over the years, he established a well-known fortune. His wealth was not myth.

He died in 1892 at the age of eighty; he was on his deathbed while in the very midst of attempting to describe to his son-in-law the location of his mine. The Sublett mine in the Guadalupes hasn't been found since.

Another tradition dates from 1867 and locates a fortune just east of the Guadalupes near Van Horn, Texas. In 1867, the Rio Grande was the refuge line for two distinct classes of defeated men: Confederate soldiers, mistrustful of carpetbaggers and fleeing to Mexico; and the entourage of the deposed Emperor Maximilian of Mexico.

Maximilian, a puppet installed by Napoleon III, was captured by revolutionaries and shot in 1867. Prior to capture, however, tradition claims that he dispatched a huge fortune to the United States for eventual transfer to Austria.

According to Dobie's interviews, Maximilian's entourage of fifteen wagons was encountered by a band of six ex-Confederates. The Americans murdered the Maximilian party and burned the bodies, wagons, and other evidence.

"Papers taken from a chest," Dobie wrote, "revealed that the leader of the dead men, an Austrian, was one of Maximilian's followers entrusted with carrying the royal fortune out of Mexico to Galveston. From Galveston it was to be shipped to Austria, where the Empress Carlotta had already gone and . . . where Maximilian wanted to flee."

The robbers, scared to be found with the chests of gold and jewels, buried the fortune and planned to return. Shortly afterward, the Confederates were attacked by Indians. Five of the Confederates were killed. The lone survivor ended up in Denton, Texas, where he died leaving, however, his story and a map of the treasure site with his doctor and his lawyer.

Another cache of loot in the Van Horn area is that of the Jim Hughes' gang.

The date was 1879, and the Wild West was in full swing. Hughes, a veteran of the Lincoln County War where Billy the Kid won fame, and his three colleagues had been followers of Johnny Ringo. They were much sought by sheriffs and had escaped into the barren area south of Van Horn, where they threw in with a band of twenty Mexican bandits long established in the region.

Both gangs were murderous. Mexican and American posses had been trying to root out the Mexican gang for half a dozen years.

The gangs joined forces and commenced to rain destruction on the neighborhood. Finally, they formed one of the grandest raids of western lore. They would sack the provincial capital of Monterrey, third largest city in Mexico.

They prepared for the expedition by raiding a U. S. Army outpost, where they obtained guns and mules. Then they mounted up and rode down to Monterrey, three hundred miles to the south.

They entered through the Monterrey gates by mingling with a caravan of guano haulers. The gangs made camp near the city walls and spent several days and nights accustoming the Mexicans to their presence.

One night, using a game of monte and free tequila as bait, they enticed the bulk of the city's night watch—twelve men—into their camp and murdered them. In teams, they then sacked the cathedral, looted the mint, and cleaned out the smelter.

They hauled for the North as fast as they could move with twenty-five loaded mules. They were pursued the next morning by Mexican cavalry, which under international agreement at the time was allowed to cross the border when in hot pursuit.

The bandits arrived at an area southeast of Van Horn, their mules nearly dead. At the night's camp, the Americans turned on their unsuspecting Mexican colleagues and shot them all dead. They buried them, and they buried the twenty-five muleloads of loot in a nearby cave.

Hughes and his gang scattered to escape the pursuing cavalry. In a few short months they were captured by various Texas authorities for other crimes and hanged. Each of them on the gallows tried to save himself by telling the story of the Monterrey raid and the site of the treasure. They were laughed at and mocked. None were believed. Only later was the fact of the raid learned in Texas and that in the various towns of Texas all the men had told the same story.

Then, too, in this same region of western Texas and southern New Mexico, there is the story of the engineer's ledge. It's protagonist, too, was named Hughes.

He was a railroad engineer who drove the four-hundred-mile run between El Paso and Del Rio. He had a mania for picking up samples of rock along the way. He kept no notes, boasting he knew every ravine, bend, cut, and canyon "like a book."

It was his custom to send promising specimens to Denver for assay. After one run, he received a report showing the ore to be so rich he would be wealthy for life.

Engineer Hughes went back to where he thought his ledge was, not far east of El Paso. He couldn't find it. He spent the next seventeen years searching for it, up and down the line, revisiting every place from which he had taken samples. East and west he searched. At first, the trainmen gave him rides. But as his old acquaintances dropped away, the newcomers did not want to bother with a crazed old man. He died penniless, a pitiful figure.

The above legends of cached loot, Maximilian's treasure, and fabulously rich lost mines are all within three to five days' horseback ride from the San Andres and Caballo mountains, where Willie Daught and Doc Noss and Bailey's clients claimed to have found gold. None of the above legends, however, has been cited as the possible source of the gold. The most commonly mentioned sources are the "Lost Padre mine" and the "Padre La Rue mine."

Those two separate myths are somewhat similar in content and may have a common historical source.

An early reference to the Lost Padre mine is found in the Galveston *Daily News* in an article dated June 22, 1873. It was transmitted from El Paso, which was then known as Franklin:

"Franklin, June 4, 1873—A short time ago two old shafts were discovered in a mountain about a mile from town and a company was formed to clear them out and prospect them. This was done and the prospectors found a well-defined silver lead, several feet in thickness variously estimated at from $30 to $75 per ton. One of the shafts is 90 feet deep and the other over 100. There is an old Mexican tradition of a very rich mine about 2½ miles from the Cathedral tower of El Paso del Nort [that is, Juarez], and this mine, the locators seem to think, fills the bill."

The 1873 newspaper refers to it as "an old Mexican tradition." Nobody to this day knows how old. Frank Dobie wrote, "the tradition of this famous mine was hoary long before guides became the chief functionaries of the church at Juarez."

The basic legend states that by standing either at the door of the church or in its towers (versions differ) exactly at sunrise and looking to the northeast, one can see the face of the mine shining in Mount Franklin.

Mount Franklin is a huge mountain of rock looking much like Gibraltar and hovering over El Paso like Gibraltar does over the Straits. Mount Franklin extends north ten miles, and part of it is within the boundaries of the White Sands missile complex.

The Padre mine was not only rich in veins, but also—as the legend has it—as a cache.

Onate, founder of New Mexico as a colony and the first governor (1598–1609), walled up in the Padre tunnel 4,336 ingots of gold, 5,000 bars of silver, 9 muleloads of jewels from Aztec treasure houses, and—possibly more valuable—4 Aztec codices smuggled from Spain.

In 1888, a miner by the name of Robinson found a shaft of "very peculiar filling." It was red soil, apparently from the Rio Grande, for there is no such soil on Franklin. He was unsuccessful, however, in digging out the filled shaft.

In 1910, an old Mexican woman gave some corroboration of the story. She told how, before she was born, a priest from Isleta, a pueblo thirteen miles south of Albuquerque, had brought Indians down to Franklin and made them fill a shaft with red river dirt.

El Paso newspapers report a steady stream of searches for the Lost Padre. A typical example was the digging of L. C. Chris, who spent years digging into what he presumed to be the Padre tunnel. Its shaft "like all other abandoned Spanish mines has been filled up by the waste and dump being thrown into it," reported the El Paso *Herald*. His continued digging began to stir up the town.

"The story of the Lost Mine in Mount Franklin that Indian and

Mexican traditions have brought down the ages to the present time, has again become a subject of much discussion. . . .

"[It] is located in the heights of Mount Franklin seven miles from main plaza. Parties have visited it by the score and find everything to indicate that this is the mine that formerly belonged to the Juarez cathedral. . . .

"Sunday morning a party of six responsible citizens of El Paso drove out to the site and thoroughly inspected the works. They found a shaft sixteen by thirty feet, sunk 125 feet in the solid rock and at the bottom found a tunnel running to the north that was sealed up with masonry. . . .

"While working in the tunnel Mr. Chris found many valuable articles which he sold for considerable money. One was a copper anvil which he brought to the Walz store on El Paso street where it can be seen yet. . . . The most valuable thing found was a horseman's spur which was made of solid silver. It was fourteen inches in diameter and long prongs projected like the spokes of a wheel. This Mr. Chris sent to the Smithsonian Institute and received $150 for it."[8]

There is no sequel to the Chris story. He vanishes, and so does the shaft.

Some thirty years later two men were reported searching for the Lost Padre mine, but in fact they were seeking the Padre La Rue mine, a different legend, which will be discussed later in detail.

Two El Paso fliers, Don Thompson and Jack Ronan, had begun a systematic aerial search for the mine above El Paso. Finding nothing promising on Mount Franklin, they kept pressing north until they arrived in the Organs, near the giant Cox ranch in Dona Ana County. There they found a promising site. It was hidden by a rock slide, however, which covered what they believed to be the mine with shale and boulders.

The next weekend they went on foot to the canyon, where to their surprise they found an aged Mexican, alone in a barren canyon in the middle of nowhere, "working at the top of the rock slide with pick and crowbar. According to Thompson, they questioned him and received noncommittal answers."[9]

---

8 El Paso *Herald,* Jan. 11, 1890; May 24, 1899; and Aug. 20, 1901.
9 Ibid., June 13, 1930.

Ronan, who spoke fluent Spanish, asked him if he was searching for the Padre's mine. "Turning white as a sheet, the Mexican broke down and admitted it was his objective."

The Mexican said he had found a cross-marked slab, similar to an old legend, but it had been moved by the rockslide from its location over the old mine entrance. After telling the story, the "Mexican wandered down the mountain and was never seen again."

Thompson and Ronan staked out mining claims and began excavations. "A crudely forged octagonal chisel, apparently of great antiquity, was the first encouraging discovery. This implement was found in a cave in the canyon wall after removal of part of the slide. It was later described by experts as being of primitive pattern such as the Spaniards favored centuries ago. Removal of hundreds of tons of rock revealed several deep caves, but not the entrance to the legendary mine."[10]

It is a tendency of myth to grow in fictional qualities as its distance in time and space increases from its source. Dona Ana County, where the pilots found the Mexican digging at the rockslide, seems to be the source of the La Rue legend. Therefore, one may assume, an examination of Dona Ana legends will bring us closer to the truth of Padre La Rue. And possibly it will bring us to the truth of White Sands.

It is in Dona Ana that Willie Daught found gold bars. In Dona Ana, too, Doc Noss found his gold. And, according to their claims, it was in this same county that Air Force captain Fiege found his and the clients of F. Lee Bailey found theirs. So, too, did Joe Newman, whose grandfather was a pioneer of Dona Ana.

Let us go then to Dona Ana to seek an early reference to a lost La Rue mine, a mine that was covered by a rockslide.

[10] Ibid.

V

The
La Rue Treasure

*. . . just before his mysterious death, Colonel Fountain exhibited
some black quartz exceedingly rich in massive gold, and stated
that it came from the heart of the Organs. . . .*
—*Griggs' History of the Mesilla Valley*

It was April 1880. In the town of Mesilla, then county seat of
Dona Ana, Sam Newman, editor of the newspaper *34*,[1] took his
seat in the adobe courthouse, pushed back his cuffs, sharpened
two pencils, and waited for the trial to begin.

The courtroom was filled. A friend of the notorious William
Bonney, Billy the Kid, had murdered a cowboy. The Kid was present that day, as was his riding partner, Pat Garrett, a horse
wrangler who had just taken a bride and family responsibilities in
adjacent Lincoln County. (Within the year the tall, six-feet, four-
inches Garrett would be elected sheriff of Lincoln County on a
campaign pledge to capture or kill the Kid. Eight months after
that he would ambush Bonney and shoot him dead in a darkened
room at Fort Sumner.)

The defense lawyer was Albert J. Fountain. The most prominent man in the territory, Fountain had been a Union Army hero
in the Civil War; then artillery commander for the revolutionary
army of Benito Juarez; and most recently president of the Texas

[1] The newspaper *34* was founded in 1878 and named in honor of a thirty-
four-vote majority won by the Democrats in the congressional election of
the district. The trial courtroom still stands in Mesilla.

state Senate. He appeared in court that April day wearing the uniform of major in the New Mexico territorial militia. Of the three men—Billy, Garrett and Fountain—Fountain was probably the most merciless, the most deadly.

Before the gavel banged, there was a commotion outside the courthouse. The trial was interrupted. Another killer was on the prowl. Victorio, with about a hundred Apache warriors in war paint and fighting regalia, had been seen passing through Silver City and headed toward Mesilla.

Fountain left immediately to muster the militia. The ranchers and cowboys in the courthouse departed for their homes. All further business was postponed.

We will attend to the Apaches and Victorio in a later chapter. They, too, play a role in the White Sands treasure. Now the focus is on Major Fountain. This important figure in New Mexico history, a man who will vanish in White Sands in a mystery yet unsolved, is a connecting link between the La Rue legend and the rockslide sighted by the fliers in 1930.

Albert Jennings Fountain was a New Yorker, born on Staten Island in 1838. While a student at Columbia College, he went on a tour of the world with a tutor and five other Columbia students. The boys deserted their tutor at The Hague, took a boat to Calcutta, and left it at Capetown to explore South Africa. They abandoned that project and boarded a vessel for Hong Kong that proved to be an opium-smuggling craft. The boys were arrested, sent to Canton, and eventually released through efforts of the American consul. They went on to San Francisco, where the party separated.

Fountain took a job as a reporter with the Sacramento *Union,* the same paper that a few years later would employ another itinerant journalist, Mark Twain.

The *Union* sent Fountain to Nicaragua to cover the 1855–56 Filibuster expedition by William Walker. Fountain was arrested by Walker, ordered to be shot, escaped in women's clothes, and made his way back to the United States.

Back in California, Fountain studied law in San Francisco and was admitted to the bar. Following the Civil War, he took advantage of the patronage being handed out by the victorious Republicans and was appointed a judge in Texas—a Reconstruction, carpetbagger judge. He was shortly thereafter elected to the Texas Senate representing thirty-two western counties and eventually became president of the Texas Senate.

In 1875 he returned to New Mexico, the scene of his Civil War action, and continued a career in politics, the law, and the military while at the same time writing florid and shameless self-promoting articles in New Mexico and Texas newspapers.

He was ambitious, vain, and very fond of his effect.

A typical Fountain story, characteristic of the man and the times, appeared in the Mesilla *Independent* and recounted a brush with Indians he had in the year of 1880. The site was near the northern end of the Jornada, the malpais, or what men of the next century would call Trinity Site.

Just as the sun was sinking we came to a wall of lava rising abruptly twenty feet or more above the plains and extending for miles in each direction. I knew the place and also knew that immediately before us was the only pass through the wall of lava for many miles. Sitting carelessly on my horse, my Henry rifle held in my right hand and resting across the pommel of my saddle, my bridle rein and lariat with which I led the mule grasped in my left, I entered the narrow pass which barely afforded room for horse and rider. The perpendicular walls on either side rose nearly to the height of my horse's head. When about half way to the summit my mule suddenly reared back, jerking my bridle hand and throwing up my horse's head. At the same instant I looked into the muzzle of a rifle not ten feet distant and behind it the head of an Indian. The rifle was discharged, the ball taking effect in the head of my horse; he fell, dead, carrying me with him. The first shot was followed by a volley and as I went down crushed and stunned under my dead horse, an arrow passed through my left shoulder, a bullet entered my left thigh, and an arrow severed the artery in my right forearm. As I lay crushed and bleeding the Indians rushed on me. The pass was so narrow that but one could approach me at a time. Lying on my back under my dead horse I fired shot after shot from my repeating rifle. I had no occasion to look through the sights as my assailants were not three yards from me. In less than a minute it was over. During that brief period I had discharged ten shots. This was perhaps the first time the Indians

*had encountered a repeating rifle and its work must have as-
tounded them. The last Indian I saw was a villainous looking ras-
cal whose only garment was a breech clout. He stands before me
now, his every feature photographed in my memory. I can see the
expression of his countenance as he came upon me with an up-
lifted lance to give the coup. He was within six feet of me when I
fired, extending my rifle in one hand as if it were a pistol, the
muzzle not more than six feet from his body, then he disappeared
and I saw him no more.*

Fountain spent the night pinned beneath his dead horse, putting
tourniquets on his wounds.

*The Indians could not reach me without exposing themselves
to my fire at short range. They knew I was wounded and desperate,
and they must have suffered severely from my fire in the first
attack. I felt satisfied they would wait for daylight to finish the
work if I were still alive.*[2]

Fountain was rescued, however, before dawn.

Fountain's second affection, after himself, was for politics, an
element of life taken quite seriously in Dona Ana County. (In one
1871 election clash between southern Democrats and northern
Republicans, there were left on the field nine dead and many
wounded.)

Political significance was attached to many minor happenings.
Fountain's apprehension and prosecution of criminals, cattle
thieves, and outlaws were frequently given political shadings. His
opponents disliked him intensely; his supporters admired him
with nearly equal passion.

His death came shortly after such a prosecution. On January
21, 1896, Fountain attended court in Lincoln County for several
days, acting as special prosecutor. He obtained from the grand
jury an indictment of cattle theft against several men, including
one Oliver M. Lee.

It created a stir because Lee, a respected rancher who was quite
tough with hands and gun, was a foremost political enemy of
Fountain. The indictment was widely denounced as a political
vendetta.

[2] *Independent* article quoted in William Keleher, *The Fabulous Frontier*
(Albuquerque: University of New Mexico Press, 1962), pp. 204–5.

When the court business was finished on January 30, Fountain started for home near Las Cruces, taking with him a number of affidavits pertaining to the cattle-stealing matters. His nine-year-old son Henry was with him. They rode home in a buckboard. Tied behind was a pony, newly purchased for Henry.

Before leaving town, they were warned by friends that there were rumors that Fountain's life might be in jeopardy. They asked him to postpone the trip for a few days until they could accompany the U. S. Mail coach for protection.

Fountain picked up a shotgun from the buggy, patted it, and said, "This will be my protection."

Two days later the man and the boy had failed to arrive at Las Cruces. A search party was sent out, and halfway between Las Cruces and Tularosa, near the public road, the buckboard was found. Boottracks of three men were plainly seen by the searchers, but no sign of Fountain or his son.

Apache trackers were brought over from the Mescalero reservation, but after following the prints of strange horses for four miles, they lost the trail.

A large search was mounted, spurred by an offer of a ten-thousand-dollar reward. Neither man nor boy was ever found.

Murder was assumed, and Fountain's enemies openly rejoiced. If it was murder, they heartily approved of it, regardless of time, place, or manner. (Sheriff Pat Garrett would later present charges in court against the chief suspect—Oliver Lee—but the charges were backed by very thin evidence, and Lee, quite correctly, was acquitted, in one of New Mexico's biggest trials.)

There were many besides Lee who had motives—cattle rustlers, politicians, gunmen—all of whom had suffered at Fountain's hands. Another motive was suggested at the time, a motive of profit. It was rumored that shortly before his death, Fountain had found a fabulously rich gold mine in the Organs. He was murdered to obtain the secret of its site.

A Las Cruces historian, L. H. Davis, a few years later reported that "just before his mysterious death, Colonel Fountain exhibited some black quartz exceedingly rich in massive gold, and stated that it came from the heart of the Organs. He claimed to have a copy of some old records found among the archives of an old church, either in Mesilla or in the mission of Dona Ana, which stated that near the old San Augustine Springs there was one of the richest mines in Spanish America, that was worked over a century ago with smelters, employing several hundred miners.

"The description was very detailed and had all the earmarks of truthfulness. In his [Fountain's] opinion, the mine and ancient camp had been destroyed by torrential floods and covered by boulders and gravel washed down the sides of the mountain. It was one of his chief ambitions to uncover what he believed to be one of the greatest gold mines in the country."[3]

It is unknown if the Las Cruces writer was aware of the 1883 report of a prospector finding ancient Spanish workings and smelters at the Modoc mine. It is probable that he wasn't, or he would have mentioned it.

The Modoc is directly behind the White Sands headquarters. San Augustine Springs is probably old Spirit Springs, likewise located directly above the headquarters. The Modoc and the Springs are part of the same peak.

Davis investigated further. He found that during the Civil War Union Colonel George W. Baylor was encamped in the north end of the Organ Mountains, either at the Cox ranch or at San Augustin Pass, which is about two miles from the Modoc. A scout came in with some white quartz containing an abundance of gold. He had found it in a canyon. "Several years later," wrote Davis, "Colonel Baylor returned and spent weeks and months in search for the milk-white quartz vein, but was unable to find it."

Those locations—Baylor's and Fountain's—center on the mountain above the White Sands headquarters, an area that used to be the Cox ranch.[4]

The location of the allegedly rich lode seemed fairly precise and, presumably, easy to verify. However, as virtually all prospectors will attest, mountains are excellent hiding places, and the par-

---

[3] "Millions in Gold Hidden In Organ Mountains," *Griggs History of the Mesilla Valley* (Las Cruces, N.M.: Bronson Printing, 1930).

[4] At its zenith in the 1890s, the ranch contained 105,000 acres, many of them the most beautiful slopes and plains in southern New Mexico. The ranch continued as a cattle producer until 1945 when the Army, through condemnation proceedings, acquired 90 per cent of the acreage. John Cox, grandson of the founder, managed to salvage for his own use a bit of the ranch, including the house in which he was born in 1895.

ticular section of the Organs above the Cox is complex, steep, and dangerous to climb. Its peaks are so inaccessible that they are a last retreat for wild mountain sheep and goats.

Although neither the lost Spanish mines nor the Fountain quartz have been found, gold has been mined in the region. Newspaper accounts of the 1880s report several times of mining "in the Texas canyon south of the [Cox] ranch."

In 1906, Davis was told another story of gold near the Modoc by an old prospector.

The prospector said he had in that same year met an old man named Teso Aguirri (or Tirso Aguirre) a descendant of Indian colonists who had entered New Mexico in 1797 or thereabouts.

Aguirre at the time was doing common labor in Las Cruces and living in a shack on the slope of the Organs. He and the prospector became friends, and Aguirre invited the prospector to use his simple mountain shack as headquarters.

At supper one night, Aguirre told the prospector a tale of how his grandfather was led into New Mexico by a renegade priest named La Rue.

By the terms of the legend, Padre La Rue had become a great man among the Indians, their benefactor and protector. In the Organs, he found a rich gold lode and mined it—not for the Spanish, but for the Indians' own benefit. La Rue and most of his Indians were later murdered by the Spanish but had hidden the mine prior to the attack.

When the story was done, Aguirre invited the prospector to accompany him to Spirit Springs. They arrived at the base of a high cliff, and at the corner of the cliff was a small room or cave.

Davis would later write, "With lighted candles they crawled in a little ways and entered a large room carved out of the rock. By the dim and flickering candlelight they saw a couch sculptured out of solid rock, and it was stained with what might have been the blood of the murdered priest." The prospector asked Aguirre where the old mine and bullion were hidden. Aguirre replied that it was nearby, but he was not permitted to tell because of the sanctity of the treasure, and of Padre La Rue, among the Indians.

Writer Davis pursued the story no farther until his interest was again aroused in 1915. In that year, some Spanish refugees from the Mexican revolution came through El Paso on their flight to Spain. They brought with them to El Paso a document found in an old church in Mexico City. It stated that in the "eastern side of Sierra de las Orgones, two days north of Paso del Norte [that is, Juarez], is a cave in which is buried a lot of bullion taken from a mine nearby. It is at a point between the Soledad Peak and another high peak to the north."

The "high peak to the north" of Soledad Peak is probably either Baldy Peak or Organ Peak, both about two miles south of the Cox ranchhouse.

Davis reported that a copy of the refugees' document was left with their friend the German consul, requesting him to investigate the region as their agent.[5]

By this time, Davis had acquired various, though circumstantial, corroborations of the Aguirre story. These ranged from conversations with Colonel Fountain, to the gold quartz found during the Civil War, to the document left with the German consul in El Paso. Accordingly, in 1917, Davis printed the legend of Padre La Rue and his lost mine.[6]

He stated that in 1796, ten young priests from France volunteered to be missionaries among the Aztecs and other Indians of Mexico. Among them was a young French noble named Father La Rue. He arrived in Mexico City in 1797.

After about six months of indoctrination and training, Father La Rue was assigned to a little pastorate in the North, located "ten days' journey" south of present-day Juarez.

"There," writes Davis, "he took up his work at a farm or hacienda, his little flock being mostly peons and poor Spaniards. Among the latter was a very old man who had been a soldier and had traveled over much of the territory then belonging to the Mexican provinces of Spain while campaigning in the army, and particularly in what is now New Mexico. A warm friendship sprang up between the old soldier and the young French priest. Finally the old man became very sick, and during his last days he was nursed and tenderly cared for by Father La Rue. Touched by this brotherly kindness, the old soldier confided to the priest a se-

[5] Mexico was undergoing a revolution in 1915, with the forces of Pancho Villa in the field against the troops of President Carranza. Many foreigners fled the country during that period.

[6] "Millions in Gold Hidden in Organ Mountains," op. cit.

cret he had carefully kept, and told him of a wonderful place of gold he had found on one of his scouting trips in former days, which he said was in a high mountain range some two days' journey northerly from Paso del Norte. 'It is called Sierra Organos,' said he. Near San Augustine Pass to the eastward, he affirmed, was a spring, which he called Spirit Spring, near which was Cuerza Vegas, cave of the meadows. A short distance from this spring was a very high rock, or cliff, and near this would be found veins and placers of gold."

Shortly after telling Father La Rue the legend, the old Spaniard died. The little colony, always on the border of starvation, then entered a season of extreme drought. Crops were meager, and famine threatened.

Father La Rue called the people together and repeated the story of the Spaniard. He asked if they were willing to gamble and go search for the place en masse. To remain meant that many would die. To go gave some hope.

They set out. We don't know how many their number, but we can assume that there were more than two hundred. They arrived, according to the legend, at a site in the Organs where, after only a few days' search, one of the men brought back nuggets of gold from a nearby canyon.

La Rue was satisfied that the old Spaniard had been truthful. He established a permanent camp. For himself, he had a room hewn out of a high chalk cliff near Spirit Spring. The other houses were made of rocks and mud mortar. The entire colony then set about looking for gold.

"Some of his men," writes Davis, "worked in the placers in the canyon, while others prospected for the leads and other sources of the placer gold. At last they found a rich vein, and selecting an obscure point where observation would be difficult to outside intruders, they tunneled into it and found it of exceeding richness. They worked the placers and the vein for several years, taking out large quantities of gold, which they deposited with Father La Rue, who acted as treasurer, who used what was necessary in purchasing provisions and tools from distant settlements. The remainder of the gold was secreted in the artificial cavern where he lived."

As the mining went on, a high stone wall was built about the mountain settlement, and no strangers were admitted. Smelters were built in the recesses of the mountains, and the camp was alive with mining activity.

Years passed. The gold bullion stacked up in vast amounts. La Rue would release only enough of it to keep the colony in supplies. Demonstration of great wealth, he feared, would bring the outside world crashing in to destroy the little colony.

He became old and revered by his people.

Meanwhile, during all the absent years, Mexico City had received no report of his whereabouts. According to Davis, an expedition was sent to find La Rue and his colony at the old site, ten days south of Juarez. Davis gave the name of the search leader as Maximo Milliano. The name does not appear in later versions of the legend, and Davis does not give his source. Presumably it was Aguirre.

Milliano arrived at the old site and after diligent inquiry learned that the colony had moved to some unknown place in the North. Milliano returned to Mexico City, and a second expedition was mounted. It was a large one, and it was charged to find La Rue no matter how far North he had gone.

The Milliano expedition spent a year in the Las Cruces region before finally locating the colony in the Organs. When they found the walled settlement, Milliano asked admittance through the gate. He was refused. Milliano then made a formal demand on Father La Rue to deliver to him, as representative of the Church in Mexico, all the gold that had been recovered and to deliver possession of the mine.

La Rue refused, asserting that the property belonged to his people; it did not belong to the Church, nor to himself.

At that point, Davis ended his story. The remainder of it does not appear in document form until 1935, when Charles Kingsley Dunham researched the legend for an article "Geology of the Organ Mountains" for the New Mexico State Bureau of Mines (Bulletin No. 11).

Dunham summarized the Davis version of the legend, then added:

"Learning from his sentries that the [Milliano] expedition was approaching, La Rue gave orders that the mine was to be covered up and the gold hidden. When the expedition arrived, he refused to divulge the secret. . . .

"By night he was murdered by a soldier attached to the expedition, and afterwards some of the colonists were tortured but the secret was never told. The mine is supposed to have been covered up by debris from the mountains. Several people have claimed to have found it. Colonel A. J. Fountain is said to have discovered it shortly before his mysterious disappearance.

"A goat herder named Tirso Aguirre, well known in the mountains about twenty years ago, was supposed to be a descendant of one of the original miners. But none of the claims has been substantiated and the search for the lost mine continues."

In filing his claim with the state of New Mexico in 1946, Doc Noss would attest that he found "the remnants of a small underground Spanish smelter comprising two five-foot bull-hide bellows, vassos, ingot molds, and mountain mahogany that had been used as fuel."

Following Noss's murder, his widow Ova would produce artifacts of a later origin. These included letters, coins, swords, and other objects that indicated a trove of bandit loot. The objects bearing printed or stamped dates had one curious common denominator: None were dated after 1880.

It wasn't until 1973 that an Albuquerque newspaperman, Howard Bryan, solved the mystery of the dates.

VI

Place Names

[*Attorney General*] *John Mitchell told me that F. Lee Bailey had a client who had an enormous amount of gold in his possession and would like to make an arrangement with the government whereby the gold could be turned over to the government without being prosecuted for holding the gold.*

—John Dean III,
former White House counsel

Rumors of vast treasure in the White Sands region have been prevalent since at least the Civil War, when Colonel Baylor's scouts found gold quartz. The rumors remained local in nature, however, for more than one hundred years. National attention came on June 25, 1973. It came simultaneously over four television networks (ABC, CBS, NBC, and PBS) in testimony by John Dean III before the Senate subcommittee investigating President Richard Nixon's involvement in the Watergate burglary and cover-up.

The senators did not question Dean about the gold. Instead they moved immediately on to matters of more political interest, more germane to the purpose of the hearings.

Following the day's recess, however, Wayne Smith, Bailey's Washington public-relations man, answered newsmen's questions. He said that Bailey's clients proposed to deliver 292 bars of gold, averaging 80 pounds each, to the Treasury Department.

He added that Bailey's clients said the gold was part of an old Aztec treasure that was hidden on the White Sands Missile Range.

Two days later, Bailey himself told reporters that there were strange doings on the part of the government at White Sands.

"My clients saw some people take two tons of gold from White Sands in two Jeeps and a truck. This was reported to the Assistant Secretary of the Treasury and no action was taken. They are not enforcing the law."

The national press descended upon New Mexico, disembarking in Albuquerque, site of the state's only major airport. From Albuquerque they scarred north to Santa Fe, the capital and repository of Land Office records, and south to White Sands for briefings by the Army.

At the time, Victorio Peak was being called Victoria Peak. The connection of the site with the Apaches a hundred years earlier was unknown. In its briefings, the Army—following the lead of New Mexico historians and other authorities—assumed that the peak had been named for some reason after Queen Victoria.

The connection to the Apaches was discovered—or rather deduced, in Sherlock Holmes fashion—by Howard Bryan of the Albuquerque *Tribune.*

Bryan, long-legged, in his fifties, and balding, is the unofficial historian and custodian of the White Sands gold stories. He has been covering them since the 1950s.

He is a "snowbird," a Northerner who came down from the Cleveland *Press* in 1948 and has spent the time since then exploring the corners of New Mexico and writing stories of what he's seen. His specialties are the places, incidents, and people not mentioned in history books. "A lot of material," he has said, "comes from talking to old pioneers. One of my first interviews was a man—106 years old then—who was in Fort Sumner when Billy the Kid was shot. . . . He said the Kid was a real psycho, would gun down a stranger riding by on the street, just out of whim."

Bryan told newsmen there might be some treasure in the San Andres, but not in vast amounts.

"Eighteen thousand bars of gold, some claim. There's never been a gold mine like that in New Mexico in ancient times or modern. That's where Bailey's story falls apart. If there's that much gold, where did it come from?"

Bryan never met Doc Noss, but he is a longtime acquaintance of Ova Noss and had done a few stories on her treasure efforts in the 1950s and early 1960s. "She's a real pistol-packing mama, that one. She says she has a way of getting to the peak without the Army knowing. I believe her. All these claimants know what's going on down there more than the military." He pulled open a deep desk drawer and took out a two-foot stack of stories and materials on the Noss treasure.

"About twelve years ago, she caught the Army digging in there. Her lawyers confronted the commanding general about it. The general denied it. Then Mrs. Noss presented evidence—name, rank, and serial numbers of the miners.

"The Army is some embarrassed about that." He pulled out a thick report. "In 1963 the University of New Mexico museum was allowed in for sixty days. They didn't find a cave. They didn't have enough time. Here's the report. You can't tell by looking, but it was censored by the Army. I later got an uncensored draft, the original draft. You know what was cut out by the Army?"

"What?" asked the visiting reporter.

"All references to Army digging. Those were the *only* things deleted."

The reporter looked through a stack of clips, then asked: "The Noss story has been kicking around for more than twenty years, but it didn't get play in the New Mexico papers until the Watergate. Why?"

"Well," Bryan said, "it was always a good story—mainly because of the persistent way Ova Noss kept devilin' the Army to let her get at it. But it was still a treasure story, and you got to understand that in New Mexico everybody is out looking for treasure. Pick any weekend and you can find a thousand or more people spread over the state looking for gold. *Any* weekend. They got their maps, they got their metal detectors, they got their divining rods.

"What they have *really got* is an excuse to have a good time outdoors. Beats bird-watching.

"The Noss story had been forgotten for ten years until John Dean mentioned it. Shortly after that, Wayne Smith—Bailey's Washington man—came to Albuquerque. He told Ralph Looney, the editor, 'You guys are missing a hell of a story. You ought to do some investigating.'

"After he left, Looney told me, 'You know as well as I do there's nothing to this story, but take all the time you need, and find out if it's true.' "

Bryan expected it would take three days. It took more than a month just to round up basic facts.

"A big question at the beginning," said Bryan, "was where does all this gold—if it exists—come from? Doc himself, and most people since then, say he found the Padre La Rue mine, but I could find no evidence that this Padre La Rue ever existed. I went to Santa Fe and saw Fray Angelico Chavez. He's the foremost Church historian in the Southwest. He said if there ever was a Padre La Rue he has never in his research come across a direct or indirect reference.

"So the question remained: Where did the treasure come from?"

After seeing Friar Chavez, Bryan's next stop was his old friend Ova Noss, living in a trailer in Clovis, New Mexico. He asked her if she had gold bars. She said no.

"That's what I been tryin' to get my hands on all these years," she told Bryan. "Goddamn Army's probably stole it all by now. Goddamn Army's been stealin' the gold all along."

Bryan asked her if she had any proof of a treasure from Victoria Peak. Ova went to a closet and came back with objects wrapped in green felt. She showed him two swords, one of them plain, one fancy; a silver goblet; a silver bowl engraved "Brazil," and a silver napkin ring engraved "Talbot Hall, November 17, 1868."

She said, "Doc also brought out several boxes of old letters from the cave. He took them to Clovis one night and read them while sitting at a table in a cafe. He got worried about all the names—potential claimants, you see?—so he burnt all them letters. But Doc told me he hadn't found one letter in that batch dated after 1880."

1880. For Howard Bryan, specialist in odd facts of the Southwest, the date meant something. It was the year the Apache chief Victorio was killed. Could there be a relation between *Victoria*

Peak and *Victorio the Apache?* Howard Bryan began to think
about place names.

There are many place names in the region surrounding Hembrillo Basin, names that activate the imagination. Canyons called
Howinhell, Lost Man, Dead Man, and the Squaw Tit. Springs
called Wild Mule and Mountain Sheep. "Hembrillo" itself is a
misnomer. The word is unknown in Spanish, and it is probably
a corruption of Membrillo, a Spanish word for quince, or quince
tree, or a tree with a small berry. Indeed, in the nineteenth century some New Mexico newspapers referred to the site as "Hembrillo," others called it "Membrillo," and still others called it
"Mimbres."

Of all the place names, the most alien is "Victoria Peak." It
doesn't fit with the crowd. Bryan attempted to track the name. He
inquired of the military. They said they assumed it had something
to do with Queen Victoria. But they had no definite knowledge of
the name's origin.

"I went to the files of early newspapers," said Bryan. "There
were many stories on Victorio's raids, and I was struck by how
many of them are near Victoria Peak. Then I hit a big one.

"On April 7, 1880, a week after Victorio had raided Clermont
and killed a number of gold miners, he was engaged by a troop of
black cavalry. It was a set-piece battle, the only such battle Victorio ever fought. That was unusual, and they fought for two days.
Apaches just didn't do that. They hit and run. The battle was over
a fresh-water spring. The troopers were desperate for water, and
the Apaches, for some reason, wouldn't give it up.

"The name of the spring? The newspapers gave it at the time:
Hembrillo. 'The Battle of Hembrillo Basin.' It was fought right at
Victoria Peak. And why did the Apaches fight so hard? I figure
they were protecting their hiding place, their headquarters."

Bryan returned to Clovis for another interview of Ova Noss.

"I asked her if she had anything else she'd found at the peak.
She showed me thirty-seven rifle casings, made for an 1873 rifle.
She also said she found two cannon balls fastened together with a
chain. The old newspaper clips said 'mountain howitzers' were
used in the Victorio campaign.

"It became convincing to me that Doc Noss found Victorio's

old hideout. The swords, the Wells Fargo boxes, the trussed-up skeletons, nothing dating later than 1880. It adds up to Victorio and the fact his band was run off before they could move any of it."

Some seventeen years earlier, in 1956, Howard Bryan had interviewed a 109-year-old man in Albuquerque. His name was Max Madrid. He had lived in southern New Mexico during the Apache wars, and he told Bryan that he remembered well having seen Victorio. Victorio, said Madrid, was a friend of his godfather, a Mexican prospector who spoke Apache.

Before his godfather died, Madrid said, he revealed that Victorio and his warriors had buried on top of a mountain all the loot they had captured from soldiers and settlers over a period of years. They had concealed the hiding place with timbers and earth.

"I asked Max specifically about Hembrillo Springs," Bryan recalled. "He had never heard the name. He knew only that his godfather had told him that Victorio had a cave with many guns and much gold and that it was 'still up there, near a big spring.'"

Apaches. There are as many misinformations about Apaches as about treasure. Not much is known about them on a formal basis. Neither archaeologists nor sociologists have penetrated their mystery. They do not know when the Apaches arrived in New Mexico, although it was probably *after* the arrival of the conquistadors. The first mention of them by the Spanish seems to be in 1642.

The alien, the outsider who has gone farthest in knowing them, is Eve Ball, a historian who lives in Ruidoso at the border of the Mescalero Apaches' retreat in the Sacramento Mountains. She has, in her forty years as a writer, become their friend, their "little sister." The Apaches have told her that in a previous incarnation she was Apache.

She is no believer in gold treasure, but we find mention of it in a book she wrote called *In the Days of Victorio*.[1] It is a gem of a book, comparing in ways to Bernal Diaz's *Conquest of Mexico* in its charm and in its unexpected information.

Most of the story is told by a man with the musical name (pro-

[1] Tucson: University of Arizona Press, 1970.

nounce it aloud) of James Kaywaykla. Kay-way-kla. He tells of
Victorio's Apaches having caves of gold and silver. He tells, too,
of a sacred mountain, of a battle at a spring, and of other things.
In the tale of the Apaches, and Hembrillo Basin, we will encounter a magnificent mural and—in connection with that—a large, illegal gold-hunting expedition into Hembrillo in 1973. It was a
surreptitious expedition that almost certainly had support from
one or more officials at White Sands.

VII

The
Mural

The name "Apache," which they take pride in, was given by an-other tribe. It derives from "Apachu," the Zuñi word for "enemy."

—*C. W. Ceram,* The First American

There is one entrance to Hembrillo Basin from the east, from the Alkali Flats above White Sands. It is the three-mile-long Hembrillo Canyon.

Passing through overhanging cliffs, it climbs steeply from the White Sands Flats, thirty-nine hundred feet above sea level, to the floor of Hembrillo, at an elevation of fifty-one hundred feet. James Kaywaykla knew it well, a place of excellent ambush.

"The entrance is not very wide, and between it and the narrow gap a short distance back is a spring. Beyond it, the walls are so close together that only one horse at a time can pass. For an enemy coming in from the basin it is a death trap."[1]

At the crest of the climb, as you clear the defile and are about to enter the basin floor, there is a clump of trees, the rock ruins of a habitation. Just behind the ruins, on a once-smooth vertical ledge, is a rock mural drawn by Indians.

According to legend, the ruin is a stone fortress hastily built by black cavalry troops in the Battle of Hembrillo Basin. Others be-

[1] Eve Ball, *In the Days of Victorio* (Tucson: University of Arizona Press, 1970).

lieve it is a sheepherder's hut built of loose stone. No one knows for sure.

The mural on the ledge just behind the ruin has no legend, and this is queer, for it is a stunning work of quality art—a superior, perhaps supreme example of American Indian rock painting.

It is large—some forty feet in length and ten feet high. It seems to date from at least two different periods. The older, underlying mural has faded colors and shows two "stick" figures, wearing skirts and pointing at smaller stick figures. In association with it are a number of designs—suns, moons, snakes, and geometrical patterns usually associated with the Mogollon culture, a people who moved into the Jornada around the time of Christ and remained there until about A.D. 1300. There is also in this ancient painting a mountain that appears to be exploding. Whether this is a representation, an eyewitness report, of the eruption of Black Peak volcano is unknown, but it is a possibility.

With imagination, one sees it as a tale of the Mogollons being led into the valley by the figures in skirts and being greeted by the exploding volcano, an omen that often had a benign interpretation for agricultural people. (The lava ash encouraged crop growth.)

Painted on the same ledge, but not obliterating the prehistoric art, is a much more vivid, newer mural. Many colors were used—black, white, red, blue, yellow—and show quite clearly a giant figure of an Indian warrior leading smaller warriors in an attack on a wagon train. It is Apache art, quite well done. It is as representational as a magazine illustration, except for one thing: Between the warriors and the wagon train is a ghostlike object, taller than the giant warrior; a figure or object enveloped in an aura of light.

Very few people have seen the mural. The 1963 exploration of Victorio Peak by the Museum of New Mexico was not allowed to enter that area of the basin. No archaeologist or geologist has ever reported its existence. But photographs have been taken of it, and it has been seen by some of the treasure hunters of the 1970s.

One of these was Joe Newman of El Paso, grandson of the *34* editor Sam Newman. "I was stunned by how beautiful it was. It belonged in a museum. I just sat down on the ground and looked at it for what seemed half an hour."

As Newman looked and engaged in aesthetic and perhaps metaphysical wonder, he discerned a more practical aspect of the mural.

He believed it to be a treasure map, with the aura of light in the center representing Victorio Peak, and other portions showing the location of a treasure.

That is, of course, speculation. What we do have, without dispute, is a glowing object set between Apache warriors and a wagon train. If the ghost represents the peak, then our mural could be showing the battle of Hembrillo Basin or, as Newman and some others believe, a map of the treasure.

But if the ghost does not represent the peak, or some other real object, then it is an abstraction, perhaps a metaphysical vision of the relationship between the Apache and the white man.

The San Andres Mountains, which play a central role in this treasure story, had a special significance for the Apaches. Salinas Peak, highest in the San Andres, was the Apache "Sacred Mountain." Since their arrival in the Southwest some four hundred years ago, Apache medicine men had gone to Salinas Peak to "obtain power." They still go there, including the present chiefs and medicine men.

The power is obtained as an adolescent and renewed thereafter. The boy, or sometimes a girl, climbs the mountain with no food or water and only a blanket for clothing. Four days and four nights are spent praying and fasting. During that time a vision may appear telling the aspirant what specific power the medicine will be. A power may be that of locating ammunition, or locating enemies, or locating water, or to cure illness. The symbol of that power, called "medicine," may be a bullet, mother-of-pearl, an herb, or just about anything. If an aspirant is given *power,* he is told in the same vision what his *medicine* will be. For the remainder of his life he carries the medicine in a pouch suspended from a cord around his neck.

The prayers on Sacred Mountain, and all Apache prayers, are sent up to their single god, Ussen.

The sun is the symbol of Ussen, and gold, being the color of the sun, is sacred to Apaches. In Apache religion, the sun is Ussen, "creator of life"; the earth is the mother "receptacle of life"; and the stars are the children. It is forbidden to dig gold from the mother.[2]

"Years ago," said Apache Ace Daklugie, son of Chief Juh, "mountain gods were angered [by the gold mining] and they danced and shook the mighty shoulders and opened up and swallowed towns. The rivers changed their courses."[3]

The Apache who will tell us a tale about gold "stacked like cordwood" is Nana. Of the nineteenth-century Mescalero chiefs, perhaps the most revered and deadly was Nana, Victorio's successor. Not as well known as Victorio, Cochise, or the others, Nana nevertheless "was our greatest chief," said James Kaywaykla. "For every white-eye killed by Victorio, Nana killed ten white-eyes in revenge for Victorio's death."

Nana thought that the white lust for gold was sacrilegious.

"The white-eyes are superstitious about gold. Their lust for it is insatiable. They lie, steal, kill, die for it. If forced to choose between it and things many times exceeding it in value, they unhesitatingly choose gold. Little do they care that they incur the wrath of the mountain gods."[4]

All the Apaches interviewed by Eve Ball regarded the white man as "superstitious" about gold. To illustrate, they referred to the gold rush that caused an invasion of Apache domain in 1879. In that rush, say the Apaches, they met an invader they resented far more than they did the trapper or the rancher.

"It was the prospectors and miners," said James Kaywaykla, "whom we considered most objectionable, for they groveled in the earth and invoked the wrath of the mountain gods by seeking gold, the metal forbidden to man. It is a symbol of the sun, of Ussen himself, and sacred to him." Apaches wear silver ornaments, not gold. They do not touch gold. "Many of my people know today where there are large deposits of it, but not one would touch it," said Kaywaykla.

Thus the Apaches have a religious tradition against gold. Yet, surprisingly, in the summer of 1976, the Mescalero Apache nation

[2] Author's interview with Eve Ball at Ruidoso, New Mexico Aug. 10, 1976. Miss Ball is the foremost authority on the customs and history of the Mescalero Apache.
[3] The reference is to an 1880 earthquake, according to Miss Ball.
[4] Ball, op. cit., p. 32.

negotiated with Joe Newman of El Paso a gold-hunting contract that stated:

"Whereas, the treasures of Chief Victorio would be the property of the Mescalero Apache Tribe . . . the Mescalero Apache Tribe hereby contracts and employs Treasure Finders Incorporated . . . to explore, excavate and retrieve alleged treasure located in the vicinity of White Sands Missile Range in the State of New Mexico. . . ."

The contract with Treasure Finders, Incorporated, a company headed by Joe Newman, was approved by the tribal council and signed by Wendell Chino, president (and chief) of the Mescalero Apache Tribe.

Chino furthermore has told the Apaches that he had been in consultation directly with President Gerald Ford in 1976 concerning the gold search. "If necessary," said Chino, "I can obtain an executive order granting us permission to enter White Sands."

Eve Ball, at an interview in her home in Ruidoso, was asked about this apparent contradiction in Apache behavior. Had they not violated their own taboo?

Miss Ball, a tiny lady with gray hair, leaned forward in her Victorian rocking chair. She smiled and said, "Chino won't get it in gold, honey. He'll get it in checks."

The present-day Mescalero Apaches are not impoverished. To the contrary, their home in the Sacramento Mountains embraces one of the most beautiful forests on the planet. The Apaches possess flowered valleys, clear streams, meadows, and stands of ponderosa pine that span 668 square miles, more than half the size of Rhode Island. They have modern farming equipment, and their fields are productive. The corn is lush, and the cattle are fat.

The affluence, however, is a phenomenon known only for the past thirty years or so. During the preceding half century the Mescaleros, cheated by federal Indian agents and encroached upon by gold-rush prospectors, were the continent's poorest people. They literally grubbed an existence out of the desert by eating the grubs of insects. Many Apaches starved or died of cold and disease.

If, as Kaywaykla and other Apaches claim, their chiefs such as

Victorio and Nana knew the whereabouts of gold, why didn't they use it?

Eve Ball makes a point that the taboo can be circumvented by the use of white men as agents. Apaches wouldn't have to dig or touch the gold themselves. Why, then, did they starve?

The answer is the elusiveness of treasure. It is an old cliché, but a truth nevertheless. Treasure has a habit of disappearing from where it is supposed to be. Men lose it.

For a tale of lost treasure, and of a brilliant Apache tactical victory at Hembrillo Basin, we go back to 1789 to meet Victorio and Nana.

On May 10, 1789, Sam Newman strode into the office of *34* newspaper in Las Cruces and laid his six-shooter on the desk. He had need of the gun. He was the most jail-prone editor in New Mexico. Inflammatory.

On this day he had been brooding over the increasing friction with the Apaches, brooding over the lack of help—indeed the very absence of *interest*—by Congress.

He puffed his cigar until it was smoking like a chimney, pawed his hair, scowled, and let his pen swoop to the attack.

"We wish it were possible for the Apaches to reach some populous point in the New England or Middle states and murder a few hundred peaceable citizens there. It might bring the people of the East to their senses. It would, at any rate, *subtract* from the number of sympathizers."

The Apache who most concerned Sam Newman was the chief, Victorio.

Victorio: To the small Apache boy James Kaywaykla, he was the symbol of his race.

At the first religious ceremony James was allowed to attend, he recalls, "I was awakened before dawn to see the erection of the ceremonial tepee [where the young virgins of the tribe would] . . . share the attributes of White Painted Woman, Mother of all Apaches, through her son, Child of the Waters.

"After skins were spread over the tepee, the chiefs joined the medicine men. A magnificent figure clad in white breech clout and moccasins, heavily beaded appeared. A scarlet blanket was thrown over his left arm; and a scarlet band held his long loose

hair in place. Victorio! I had never seen my chief dressed otherwise than his men, and did not recognize him at first. I had heard
Grandmother speak of that blanket—the gift of Manuelito, Chief
of the Navajos. Victorio was not as tall as Naiche, but I think he
was the most nearly perfect human being I have ever seen. No
Greek statue had greater majesty of form and bearing.

"Behind him came Grandfather, similarly attired; then the
others. I realized that my people were beautiful. I had never
thought of them so before. Dimly I sensed pride in my race and
heritage."[5]

Also present at the ceremony was Victorio's maiden sister,
Lozen. She is the most evocative and mystical figure in Apache
folklore.

She had no husband, though she was a legendary beauty. Many
men courted her, but she refused all, and Victorio did not compel
her to choose. She did not marry because she had seen The Gray
Ghost.

"When she was very young," recalls Kaywaykla, "a strange
thing happened. Warriors reported seeing one they called The
Gray Ghost in our mountains. This one rode alone, and could
never be approached. He was of great stature and very powerful.
Once three warriors up on a point saw him pursued by cavalry
into a canyon. They called to him and pointed out a secret hiding
place. Gray Ghost nodded, rode into it, and stayed until the soldiers had given up the chase. He came to our camp and remained
for some time, visiting with our people. He learned some of our
words and told Victorio he was a chief from far toward the rising
sun.

"A strange wagon came through our land. Twelve men rode beside it as guards. There was also a driver, and with them an old
woman. All spoke the language of Mexico, but they were not
Mexicans. Inside the wagon they carried a young woman—very
beautiful. When they moved west, Gray Ghost followed."[6]

[5] Ball, op. cit., p. 41. Kaywaykla presumably had seen Grecian sculpture
on a visit to New York City and the museums there.

[6] Ball, op. cit., p. 14. There are several legends of a coach, guarded by a
patrol and carrying a beautiful non-Mexican woman, passing through the
White Sands region about 1868, the time of Lozen's adolescence. The date
corresponds to the overthrow of Maximilian in Mexico and the flight of the
French and the Austrians.

Lozen was too young then for marriage, but she had seen Gray Ghost, and no man ever afterward interested her. Instead, she put marriage aside and rode beside Victorio as a warrior. Her fame extended far into other tribes. Said a Seneca chief, "She is sacred, even as White Painted Woman. She is respected above all living women."[7]

Lozen's power was her ability to detect the enemy. She would raise her arms and slowly turn as she sang a prayer to Ussen. She would then feel the presence, or lack of it, of an enemy. She would know their location. According to the Apaches, she never failed.

She was also a fine warrior. Although Apache women were held in nearly equal esteem to the men and fought and rode as warriors, Lozen had a special status: She, alone of the women, was invited to the tribal council. She could ride, shoot, and fight like a man and, according to Kaywaykla, "she had more ability in planning military strategy than did even Victorio."

The events that set Victorio, Lozen, and the Mescalero Apaches on the warpath in 1879 had their beginning nine years earlier, in 1870, when the Apache tribes were given a reservation at Ojo Caliente in central New Mexico.

The site, rich in land, water, and natural beauty, was part of an ancestral range and much prized. Victorio and Cochise had been promised that they would be given food and blankets and that the land was theirs, in the words of the old Indian contracts, "as long as the mountains shall stand and the rivers run."

In 1878, most of the Apaches were summarily transferred from Ojo Caliente to San Carlos reservation in Arizona. This Arizona site was a death hole of disease and starvation and was presided over by notoriously corrupt Indian agents.

Many Apaches refused to go to San Carlos. Geronimo and some other leaders were taken there in chains. An attempt was made to do likewise to Victorio and Nana, but they and their followers escaped. In May of 1879 they set out on the warpath. Their raids began in the Organ Mountains and extended across the Rio Grande into the Hillsborough mining district, where they killed miners and stole horses and cattle.

[7] Ibid.

It was at that point that editor Sam Newman took up his pen and addressed an editorial to General Edward Hatch, then a colonel and the commanding officer of the Army in New Mexico.

"The Apaches are encamped," wrote Newman, "in an area of the San Andres between Salinas Peak and Membrillo Spring. . . . We urge that the proper steps be taken to effect their capture."

Hatch responded slowly, and by the time the troops arrived in the San Andres a week later, the Apaches were gone.

"It seems," wrote Newman, "that the Army has orders not to find the Indians."

This would be the pattern throughout. The Army could not pin, could not find, could not defeat Victorio.

In August, Newman reported that Victorio had "killed five soldiers at Hot Springs and ran off the government herd of horses, then proceeded to Hillsboro and McEvers' ranch, where seventeen more persons, including women and children, fell victims to his murderous rifles and houses and property were consumed by his fire brands. . . . As far as ascertained, over eighty of the best people in the land have thus far been murdered in this and Grant counties." Newman added sarcastically:

"During all this what has General Hatch done? Captain Beyer, notoriously unfit to command any body of troops in the field, was sent out."

The months passed, with the Apaches stacking up a series of victorious skirmishes.

The most humiliating of the Army defeats came on April 7, 1880, at Hembrillo Basin. This was the fight that caused Colonel Fountain to summon his militia and postpone the trial at Las Cruces.

Scouts had reported Victorio to be encamped at the base of Sacred Mountain. General Hatch immediately dispatched four separate cavalry troops, including Fountain's militia, to surround the Apaches. Meanwhile, a black cavalry unit in the immediate area was ordered to trail the Apaches and keep them under observation until the main force arrived.

We have eyewitness accounts of the ensuing events given by both sides—by soldiers present and by James Kaywaykla.

The Apaches had encamped at the base of Sacred Mountain. But they were not led by Victorio; they were led by Nana, who was attempting to bring a dozen warriors and some fifty women and children to join Victorio in Mexico. Among the children was Kaywaykla.

The spring at which they camped had clear, blue, and cool water. But it was not fresh water; it was gypsum water, and in that fact Nana was laying a trap to buy his escape from the closely pursuing black cavalrymen.

Kaywaykla says Nana "had forbidden us to touch the water," and as the Apaches left the spring at night, the cavalry moved in on it in the morning. "The cavalry drank the water, and so had their horses. Both had become ill from a laxative effect, and were weakened until they could hardly travel. We had not poisoned that spring; the illness was caused by a natural mineral that this one time operated in our favor."

Gypsum is a mineral similar in effect to Epsom salts.

Thomas Cruse, part of Hatch's command, described the same event:

"Pursuant to orders, Carroll [commanding officer of the cavalry troop] camped on the evening of April 5 at the Malpais Spring —which flowed water beautifully clear and cool, but dangerously charged with gypsum. As a result of watering here, nearly all of Carroll's horses and half his men were deathly ill before morning. So he hastily broke camp and moved into the mountains. He expected to find a spring where he had camped the fall before, while on a scout against these same Indians, but when he reached the locality, not a drop of water was found there.

"With much difficulty he moved southward, to reach a cañon where he was assured there was plenty of water and, at about six o'clock, totally exhausted, he entered Memtrillo [sic] Cañon, where the water was. But in this very cañon was Victorio's camp! Memtrillo Cañon was the object of our entire expedition and the concentrated attack scheduled for the next day."[8]

The Apaches moved into Hembrillo Basin, with the warriors being stationed at ledges guarding the narrow canyon entrances.

Meanwhile, Hatch's troops attempted to close the circle.

Cruse reported that "early the next morning, the Indians took the aggressive and surrounded the command, shooting into it from every direction and especially covering the water." In the mean-

[8] *Apache Days and After* (Caldwell, Ida.: Caxton Printers, 1941).

time, one of the Hatch companies—McLellan's—marched in from Aleman and joined Cruse's unit. Cruse climbed a steep bluff and "from the edge of this, as the mists cleared, we could see a pretty valley, surrounded by rough peaks with three deep cañons converging into it. Our guides said we were at Memtrillo Cañon, the location of Victorio's Camp."

By this time, the Apaches had the black cavalry unit pinned under heavy fire and blocking them from water supply. Cruse formed a relief column.

"We formed for attack and . . . drove the Indians back. Now water could be obtained, if yet under heavy fire from peaks and bluffs overlooking the spring. Gatewood and others were much puzzled at the large number of hostiles present, as it was well known that Victorio never at any time had over seventy-five warriors, while here we found at least two hundred ranged around us."

When more of Hatch's troops arrived, Cruse prepared a final charge up the canyon and into Apache lines. "I passed the word along that at the command the advance was to be continued on the run, the men firing at will until the hostile line was reached. There would be no halts for any purpose."

The charge was a failure. The Apaches had vanished from the basin. They had slipped out during the night, going over the west saddle of Hembrillo Basin.

Cruse later learned that the fleeing Apaches had almost run head-on into General Hatch himself.

Hatch had been proceeding up the Jornada and was within a mile of Hembrillo on the west side when his couriers arrived telling him that the black cavalry was pinned. Wrote Cruse:

"Here, once more, fate played against [Hatch]; instead of proceeding about a mile farther and coming up the trail in the mountains as he had intended, he turned back to regain the Jornada," thinking to travel more quickly to the rescue of the trapped cavalry.

"He reached Memtrillo about noon—after all the hostiles had left. Had he chosen the trail he would have met Victorio with all his warriors, women and children, the band short of ammunition. It is probable the campaign would have ended right there. A reconnaissance some days later showed the two trails only a short distance apart and parallel, Victorio's party going south and General Hatch's command going north, both moving rapidly."[9]

Kaywaykla reported that there were never more than a dozen braves, but the shooting seemed more intense because rifles were being fired by women, including Kaywaykla's mother and grandmother.

"The troops were easily beaten back until more cavalry came in from the Tularosa Basin. Nana took the women and children up the arroyo and around a point to the Jornada."

The trail was very rocky, and there was little dust to warn the Apaches of the coming of an enemy until they were nearly face-to-face. They heard, however, the approach of some scouts and, with the hoofs of the horses shod with cloth and the children holding the noses of dogs so they wouldn't bark, Nana hid the band against the wall of the arroyo while the scouts passed above.

The Indian scouts passed less than five hundred yards away. "It was not long before the Blue Coats came into sight. It was a very large detachment, several companies. We learned later that it was commanded by Colonel Hatch, and that he had been sent to cut off our retreat. Nana did not wait long in hiding. . . . Over the Jornada we rushed, that trail which my people had made a terror to invaders."[10]

Nana and his people joined up with Victorio, Geronimo, Juh, and other Apaches in Mexico. There, raiding to obtain ammunition and food, their flight to survive continued. As the months passed, the ammunition dwindled.

In the summer, Nana fell upon a mule train using a smugglers' trail across the border. Little Kaywaykla was excited. "Mules! It was a mule train laden with great bags of heavy stuff—ammunition, perhaps." Nana ambushed the smugglers, turned the mules around, and led them into Mexico without inspecting the bags. At nighttime they stopped, and the Apaches opened the saddlebags.

"I will never forget the disgust on Grandfather's face," wrote Kaywaykla. Nana drew a heavy dull bar from the bag and threw it to the ground.

"Silver!" he said contemptuously.

9 Ibid., pp. 73–76.
10 Ball, op. cit., p. 85.

The bar was about three feet long and six inches in diameter. One of the Apaches said the silver could be taken to Casa Grandes and exchanged for ammunition with the Mexicans.

Nana was not prepared to enter Casa Grandes at that time, and instead the silver was buried on the spot. Nana marked the spot with his eye so as to remember the cache.

Nana missed nothing. He told one of the subchiefs, Kaytennae, that evening by the fire that he knew of several places where either gold or silver was to be found in abundance: "There is a canyon in the mountains west of Ojo Caliente—a long way West—where chunks of the yellow stuff as large as grains of corn can be picked up if I did not fear the wrath of Ussen. I know of cliffs with layers of silver so soft that it can be cut with a knife. I know a cave where bars of gold are stacked as is firewood by the soldiers."

All the Apaches, said Kaywaykla, moved closer. Nana had never mentioned gold to them before. They listened eagerly as Nana described another cave of gold he had seen three days' walk from Casa Grandes:

"Just beneath the rim of a cliff I found a cave almost filled with it. It could not be reached from above, so the Mexicans must have carried those bars up ladders from one ledge to another to hide it."

Shortly afterward, Victorio was ambushed and killed. Nana and his band were among the few survivors of the ambush and attempted to head back to the San Andres. The countryside was alive with Mexican and American troops. Ammunition was short. They headed toward the cache of silver they had made a few months previously. Desperate for ammunition, they intended to buy it.

Nana was still reluctant to do it, and he let other Apaches—expert trackers all—lead the way. They repeated what had been and what would be the experience of white men and treasure. They lost it. They could not find the cache.

"I think," wrote Kaywaykla, "we must have been at that place more than a week before the young men gave up the search. Grandfather waited patiently, and he betrayed no satisfaction at their failure. Nevertheless, I think he was pleased."

Nana never spoke again to Kaywaykla of gold or silver, and

when he died some twenty years later, the secrets of Apache gold died with him.

In modern times, white men such as Willie Daught, Doc Noss, and Captain Fiege would speak, as had Nana, of seeing caves "where bars of gold are stacked as is firewood by the soldiers." A postman at Tularosa, Harvey Snow, would claim a silver vein with silver "so soft that it can be cut with a knife." He found it at White Sands.

Meanwhile, during Thanksgiving weekend, November 24–26, 1973, the most evocative of the Apache treasures—the mural—was visited by members of the F. Lee Bailey treasure-hunting group. The visit seems almost certainly to have been arranged with the collusion of one or more officials of the White Sands Missile Range.

At the time, the basin was patrolled constantly by military police in Jeeps, by range riders in four-wheel-drive trucks and on horseback, and overflown twice daily and twice nightly by helicopters equipped with infrared and radar devices. Furthermore, the two roads leading into Hembrillo were sealed by locked iron gates.

For a three- or four-day period, however, all these security precautions were suspended. Sometime Thursday or Friday, "a party of five to nine personnel with two or three vehicles, one with a small trailer" entered the basin undetected. They had keys to the west gate.[11] They drove to an old arroyo. There they made camp for three days. They made numerous dynamite blasts to excavate different cliff faces. Like the caravan itself, these blasts also were undetected.

At the conclusion of the trip, the leader of the expedition walked half a mile up the arroyo, away from the excavations. He stood before the mural. He set down a bag, took out a camera, and shot a roll of color film. He studied the mural as if it were a map. He believed it to be a map, and he wanted no others to see it. Then he took out a hand drill and bored six holes at the base of the mural. They were placed very carefully.

[11] Confidential "Security Office Briefing On Hembrillo Basin-Gold, March 5, 1974." The document is in the possession of the judge advocate general's office, White Sands headquarters.

The dynamite was next. He pushed it into the holes, snapped on the caps, and strung out the fuses. He lit his match and retreated down the arroyo to await the explosion.

It blew. The mural was gone forever.

VIII

The
Maltese Falcon

"Their favorite thing that day was to tie me up and wire my fingers to the spark plugs of the car while the engine was running. I had to give them the location of the gold."
—*Willie Douthit (Nov. 1976)*

"Willie is the epitome of a man who has found something, who has been scared all his life, who has done something wrong and doesn't want to go near the place that has anything to do with it. Hollywood couldn't have cast a better character. He is right out of The Maltese Falcon."
—*Norman Scott (Nov. 1976)*

He was known in New Mexico as Willie Daught or Willie Doughit. In California, where he has lived since 1932, he is known under a third name, Lawrence F., a name he continues to use. His birth-certificate name is William Douthit, born in Midland, Texas, in 1908.

For 2½ years—from November 1929 until the spring of 1932, ironically the lowest years of the worldwide Great Depression—he had possession of 56 tons of gold bullion worth a minimum of $32 million on the prevailing market. The gold was stolen from him, and he vanished into anonymity without filing the least criminal charge or making a public complaint.

His conduct has aroused suspicions. But the suspicions do not center around whether he truly had the gold. No, the witnesses to its existence and his possession of it are too numerous. The suspicion centers around what was in Willie Douthit's cave *in addition to* the gold.

Beginning in 1929, Douthit has told contradictory stories of how he first found the gold. His first tale, given to the Dona Ana sheriff's office, said he had obtained a map from a monastery near Mexico City. In the same year, he told a friend named Milt Holden that he had found a map while wrecking an adobe house near Rincon. Later he told New Mexico writer Xanthus Carson that he had obtained a chart from a church in Chihuahua; and in 1976 he told this writer that he obtained navigational directions from two priests who regularly visited Dona Ana County from Mexico.

Willie Douthit is concealing the true means by which he found the location of the treasure. There are rumors, however, that the discovery had sinister aspects. For instance, in 1929, Willie told a confederate that he had found in the cave "five skeletons . . . and a man in there who hadn't been dead a long time."

Nothing is proved, nothing is accused. But *if* Douthit had a connection with the dead man, an incriminating connection, then it would explain his silence, his meek acceptance, in giving up the gold to his kidnapers.

Willie Douthit moved as a boy with his family to southern New Mexico. "I was raised in that country," he recalled in 1976, "and I got to know these priests who came up regularly from Mexico. I helped them, carrying things and the like. They were looking for treasure, you know, in the Caballos. They eventually gave up. I don't remember when, but I must have been about fifteen years old. They gave me a map—or more exactly a description of latitude and longitude. They said if I found the treasure I could keep half, but I must give the other half to them."

The gold, said Douthit, was the Padre La Rue treasure. One of the priests' parishioners had found the ancient diggings in the Organs. "It was placer gold, which had been melted down and molded into bricks and stored in a cave. That was why the priests couldn't find it. The cave was apart from the smelter, and the location was lost."

In all of Douthit's versions, he states he used a compass and surveyor instruments to find the cave. He was age nineteen, small, wiry of build, and red-haired. He obtained a grubstake from a man in Hot Springs and set out for a three-months-long search. Then, he says, he found it.

What he found first was an ancient smelter atop one of the Caballos. "There was lots of wood up there," he remembers. Down below, a half a mile or so distant, he found the cave. Its entrance was small, so well concealed that entry depended upon knowing which rock to remove. From the entrance, the cave sloped down for about fifteen feet, then leveled out.

Douthit first found an old-time forge and beside it some "queer-looking hammers and quite a pile of red clay."

Behind the forge lay the bare-boned skeletons of five people, and behind them the wooden roof supports for a room. There were the remains of a door in one corner, indicating a second room. But the roof had partially caved in at that point, and Willie did not dare to enter it. Wind blew from the cave-in, and Willie had difficulty keeping his candle lit. Then, on the right side of the room, he found bars of metal. The bars were stacked "like cordwood" in a pile roughly the size of a pickup truck. They were corroded, "they looked like an old corroded bullet," and Willie had to pry the bars apart.

It was gold, and the boy sat down and counted 1,465 bars.

"When I found the gold, like an idiot, I ran down and told everybody."

The day was November 6, 1929.

Willie had been staying at the ranch of Milton Holden, located on the west bank of the Rio Grande, across from the Caballos.

Holden had grubstaked the boy and was donating him room and board. He had two sons, Joe Buck and John Holden, who were about the same age as Willie. They remembered well the day of November 6.[1]

Douthit (the Holdens knew him as "Daught") left each morning after chores for the Caballos. Willie freely admitted that he was seeking treasure, and smiled easily when he was teased about his equipment, a maritime navigation instrument seldom used in the New Mexico desert. It was a sextant.

Willie had put the story on them that he had found an old map

[1] Xanthus Carson, "The Gold Is Mine," *Treasure* (Nov. 1974).

while wrecking an adobe house in Rincon, and he had told other people at the same time that he had found a map while digging in the ruins of a nearby Army fort.

The map had been done by a sailor who had come up the Rio Grande and plotted the treasure site in navigational terms.

On November 5, Willie returned to the ranch late in the day and asked Milt Holden for a horse, a six-gun, and a hacksaw. Willis then went to a Hot Springs store and came back with a can of black pepper and some cartridges for the revolver.

Willie was up earlier than usual the next morning, cooked his own breakfast, and left for the Caballos on horseback. The Holdens paid little attention. It seemed to be just another day.

Willie returned some nine hours later, about three o'clock in the afternoon. The Holden boys and some ranch hands were, as it happened, standing in the yard. Willie rode up and very silently tossed at their feet a corroded bar of metal.

"What do you think of this?" he asked, and got down from the horse. The men started scraping at the bar. "If it's gold," said Willie, "then I have a pile as big as that old truck standing over there."

One of the men ran into the house and got a hammer. Another picked up the bar and carried it to an anvil in the yard. They hammered it and tested it in various ways, including driving nails through it to see if it was pure. When the tests were concluded, the men cheered. Willie had found gold.

And while the tests were going on, Willie handed Milt Holden the six-gun. One shot had been fired from it.

The men in the yard, including the Holden boys, abandoned all other business and started backtracking Willie's trail into the Caballos. They wanted to find his stash.

Willie's tracks, however, vanished as soon as he entered the rocky country. In case anyone tried to backtrack him with dogs, he had spread black pepper across the trail. The empty can was found on the rocky mesa where Willie's tracks vanished.

When the Holden boys returned to the ranch, Willie asked them to drive him to Hatch. He intended to spend the night in a hotel, to celebrate, and to continue the next day to El Paso, where he could sell his bar.

The Holdens cranked up their ancient Model T flivver and

drove Willie to Hatch, some twenty miles to the south. They left Willie, and his gold bar, in a room at the Jornada Hotel, had a couple of beers, and later returned to the ranch.

The following morning their ranch was invaded by more than five hundred people who had heard the news of Willie's gold and were spreading out to find it. Meanwhile, more sinister men had visited Willie in his hotel room.

"Around midnight," stated the Las Cruces sheriff's report, "three armed men entered his room, forcing him at the point of a drawn revolver to accompany them to where they had a car. He was bound and placed in the car. They then drove to the Springs crossing the river at Elephant Butte dam, by Engle to Cutter [that is, south], and into the Caballo Mountains from the east. Here they went into camp and were joined by three other men. They demanded that Doughit [sic] show them the location of the gold, but he told them he could not do so on account of not having his instruments."

Willie stated he was held at the camp in the Caballos for several days. Finally the kidnapers put Willie in a car, drove to the outskirts of Socorro, and released him. Willie's gold bar, however, they kept.

"They were just trying to scare me," Douthit would later recall. "The next time I was a little more careful. I managed to get two bars from the cave down to El Paso. I put one in the bank and gave the other to the church."

Willie was finding it increasingly difficult to visit his cave. The Caballos were alive with treasure hunters, and Willie's steps were dogged each time he entered the area. Roads into the Caballo were blocked by treasure hunters, and each vehicle was inspected for Willie's presence. Willie took to traveling only at night, and he bought a donkey.

On his third gold run, he was captured again, this time by a more serious bunch.

"They ambushed me while I was going to the cave, so they got no gold—just my poor body to work over." The group, of about ten men, included at least two deputy sheriffs of the region, men known to Willie. For two days they held him in the Caballos, tied to a tree, whipped with belts, punched, kicked, and burned with cigarettes. Willie nevertheless refused to give up the location of the cave.

Had he told, the men might have killed him. He could identify several of them, and his body bore the evidence of their treatment. And indeed he did withstand their tortures and assaults with such determination and stubbornness that it seemed his very life depended upon his keeping the secret.

After the men let him go, he went to the sheriff of Las Cruces and filed a complaint on November 19. He did not name the deputies in his complaint, but he did tell the sheriff off the record, and Willie would not again be kidnaped by law-enforcement officers.

And the sheriff gave Willie some advice: Stay away from the Caballos for a while. Let the situation cool down. Let the word get about that the law will protect the person and property of Willie Daught.

Willie went to his uncle's farm near Albuquerque. There he met Robert Ward, known as "Buster" Ward, a rowdy young rodeo performer whose father had been an early settler on the Caballos' eastern slope, in the very region of Willie's cave.

Willie and Buster talked about treasure, and as their friendship grew, Willie told of what he had discovered. This proved risky to the friendship, for as soon as Buster pieced together enough clues, he took off for the site—to the eastern Caballos, where he'd been born.

He found the cave almost immediately.

It is uncertain whether Buster intended to betray Willie. They would very quickly become partners in the treasure and resume their friendship. In the meantime, however, Buster was at the cave and, being unknown, he had much more freedom of movement than Willie.

Using a truck, he extracted six bars of gold and drove them to Douglas, Arizona, where he sold them to smelter officials.

He made at least one more trip between the cave and Douglas. On the third trip, he was found out. He was ambushed in the Caballos, tied to a post, and burned with matches. His father, Robert Ward, Sr., was told the story by Buster and Willie.

"Though they had my son tied up, as soon as he got a chance he kicked one of the men so hard in the stomach that it knocked

him out, and at that time the other fellow drew a gun and shot at
Buster. The men left him for dead."[2]

Though wounded, Buster worked himself loose and made it to
his sister's home in Las Palomas.

Following the incident, Willie and Buster formed a partnership,
with Buster, for a third share, acting as Willie's bodyguard and
helper.

They decided that their next haul would be a big one. They
would bring out sufficient bars so they could retire for quite a
while. They would have enough money to leave the country for
several years. People would have time to forget about them before
they returned.

"We took our mules into the mountains," said Willie, "and
loaded up eleven bars. We worked these down to Hot Springs,
where we put them in the bank overnight. Before the bank would
let us put it in the vault they insisted on assaying the bars. It assayed out at 66 per cent pure gold. Once we got it put away we
went to Judge Doan and asked for the loan of his car. He agreed,
and he arranged so that we could get the gold out of the bank at
first light in the morning and get away before the town was up.
We had made other arrangements to sell that gold in Lordsburg,
once we got it there.

The judge would later state that he witnessed the loading of the
bars into his car. Willie and Buster left town with the bars locked
in the tool box of the judge's car. They took back roads and
headed southwest, toward Lordsburg. The date was May 8, 1930.

"Just about a mile north of Deming," said Willie, "on Highway
26, we saw some trees across the road and a car also blocking the
highway. These were not cowboys or prospectors. These were a
different bunch altogether—about six of them, and they were like
city men. They told us, 'You youngsters are playing a men's game
now.' They got us out at gunpoint. They broke open the tool box
of the judge's car and they stole our gold."

Adjacent to Highway 26 runs the Atchison, Topeka, & Santa Fe
Railroad tracks. As the men were unloading the gold, and as
Willie and Buster awaited their fate, a freight train roared by. The
boys broke and ran for it.

[2] Ibid.

Said Buster Ward's father, "Willie jumped for the train, and so did Buster. They both caught their train, and had it not been that Buster's foot swung around and caught on a switch as he passed, he might not have happened to this accident. Instead, the next thing he knew, both legs were severed. When the ambulance arrived, my boy was propped on his elbow smoking a cigarette."

Willie Douthit had applied life-saving tourniquets, and he accompanied Buster to the Masonic Hospital in El Paso.

The treasure hunting ceased for several months while Ward was fitted for artificial limbs and practiced using them.

In late 1931, both boys returned to Hot Springs and readied themselves for another gold run. They had not been very successful so far in exploiting their treasure, and their lessons had been fearfully expensive. They had, from Ward's previous gold sales, a stake of money. They were content to bide their time while they worked out the perfect plan.

One day in December, when the Caballos were experiencing a rare snow, the pair split up to explore arroyos that might be used by trucks. Buster headed south; Willie went to the north.

And here, as Willie drove slowly through the arroyos of the Caballos, he was ambushed again. It was by the same men, "the city men" who had hijacked him outside of Deming.

"I tell you, those were real serious fellows. They told me again, 'Boy, you're playing a man's game.' They had caught up with me in the mountains. Their favorite thing that day was to tie me up and wire my fingers to the spark plugs of the car while the engine was running. Man, I tell you that does get exciting. I couldn't stand it. I had to give them the location of the gold."

Leaving Willie's car at the ambush site, the men put Willie in their car. He took them to his cave.

And here Willie's narrative stops. He has never said what the men found in the cave. He refuses to say, other than "They found about fourteen hundred gold bars." He refuses to comment on the men's reaction. He says only, "They talked about moving the gold out of the cave. They said they'd do it in relays by truck. They said they'd take it East, and I gathered they were talking about either the San Andres or Fort Bliss." Willie was released at the cave site. He walked back to his car unmolested.

There is a strong local legend in the region that in January

1932, preparations were made in Hot Springs for the rental of ten big trucks at a price of one thousand dollars.

"It was the night of January 25, 1932," said Willie, "that they took the trucks up there and moved the gold."

Willie was not in New Mexico when the gold was moved. He had left weeks earlier, almost immediately after he had been released. He and Buster split up their partnership money. Buster went to Arizona, where he died a natural death a short time later. Willie went to California and to his new life under a new name.

In 1976, Willie was found.

The man who tracked him down was Norman Scott. At a muscular sixty-eight years old, a professional diver and treasure hunter, he has participated in several treasure finds, including the excavation of the Mayas' sacred well at Chichen Itza.

Scott visited Willie in California and found him living in modest but not suffering circumstances. He asked Willie why he had never reported the theft of his gold. Willie said only, "It would have done no good."

"I found Willie's story believable for the most part," said Scott later.[3] "He's a smart, cagey guy, but he's scared of something. I met him at a public restaurant, downtown, and when he sat down I saw he had a gun—a revolver—inside his coat.

"He did not explain why he just disappeared off the face of the earth in 1932. Willie is the epitome of a man who has found something, who has been scared all his life, who has done something wrong and doesn't want to go near the place that has anything to do with it."

Why did Willie make no complaint? Why did he meekly leave New Mexico? What happened to the ambushers? What happened to the gold?

Willie Douthit provides no convincing answers. He did not report the kidnaping because "It would have done no good." The men did not sell the gold immediately "because it was stolen property so they would have been caught at it, and after that, in June 1933, it became illegal to sell gold." What happened to the men? Willie says that in the midthirties, three of them killed each other

[3] Personal interview.

in a gunfight. The remaining three died in various ways, the last man dying in 1974.

Significantly, it was only after the last man to have seen Willie's cave died in 1974 that he returned to Hot Springs, now named Truth or Consequences. It was his first visit since 1932. He came to check on his gold. And he found it was gone.

To provide some possible answers to these mysteries, let us use some deduction.

The facts are:

1. Willie had gold. This is attested to by numerous witnesses.
2. He was kidnaped, but he filed no criminal charges, nor did he make any public complaint. He quietly vanished.

Now let us hypothesize that the cave contained not only gold but a secret incriminatory to Willie; remember the "man in there who hadn't been dead a long time" and the missing bullet from Willie's gun? These would explain his reluctance to file charges against anyone who had visited the cave. It can explain why, after being robbed of $32 million, he meekly leaves the area. He was not afraid of further violence from the men. They had gladly turned him loose—once they saw the cave.

Now, what would six men apparently from some nearby city— perhaps Albuquerque or El Paso—do with fourteen hundred bars of gold that at eighty pounds a bar, weighed fifty-six tons? Apparently they did not fear complaint from Douthit, nor interference from local law enforcement. But they would fear community scandal. If they returned to Albuquerque or El Paso or wherever with gold bars, people would know it was stolen from Willie. His story of the earlier kidnaping had been in the papers. Willie was already becoming a local legend.

Not only would their families be shamed, but undoubtedly there would ensue an investigation from higher authorities, from state or federal agencies.

The consequences, the risks would be the same if they attempted to sell the gold locally or in Arizona or in other out-of-state markets. And if they tried to truck fifty-six tons of gold bars to East Coast banks, the news would surely get out.

There was, however, one possible solution, a solution common to the 1930s: Sell the gold in Mexico without recording the transaction.

This then was probably their plan. First, they wanted to forestall any accidental discovery of the cache (the general area of Willie's find in the Caballos was well known). To do this they would move the gold. A logical site was the deserted San Andres, but not necessarily Hembrillo Basin. From there they could shift it slowly, bar by bar over the years, into Mexico.

It possibly would have worked had it not been for the federal law passed in 1933 forbidding the ownership of gold. This act greatly complicated their plans. First, it was intimidating because they would be in violation not of local law but of federal law. Second, passage of the act caused a rush of gold smuggling into Mexico and a consequent increase in border patrol surveillance.

The plan still might have worked anyway had there not arrived in Hot Springs that same year the man named Noss. Tall, dark and threatening, he was also a crack shot.

IX

Hoarding Gold

"When we lived at Engle about 1936 . . . Doc Noss asked me if I knew of any caves in the San Andres Mountains and asked me to show him and I did. . . . My dad and I and my brother and most of the cowboys had been in the cave and there was nothing in there and never was."

—*Mart Gilmore, rancher*

Gold is basically different from money. To be owned, gold must be in one's physical possession. Money, on the other hand need not be physically possessed. Money is an abstraction. A person can *own* millions of dollars, pounds, francs, or marks without touching or seeing any of it. Money is arithmetic figures backed by an agreement that the figures are a medium of exchange for property and services. Money is a scorecard of commercial performance—for individuals, for groups, for corporations, and for nations.

For his or her labor, the worker receives a check, a piece of paper with figures on it. The check is placed in a bank, and the worker is given another piece of paper with figures showing his economic points of the moment. The worker in turn writes figures on other pieces of paper, subtracting from the points, to buy food, or to make payments on the house. If a worker has a lot of points he may buy a company, hire people, influence government, experience the wielding of power.

Gold, being physical and not an abstraction, is not so versatile. It can, of course, be exchanged or even substituted for money, as can any valuable object. But to own gold is cumbersome.

Put at its simplest, to own gold one must have the power to touch it, and a corollary power to protect it from being touched by others. Gold is a material reality, and this fact is the reason gold long ago ceased to be synonymous with money in commercial nations. To meet expanding and increasingly complex economies, commercial nations required a rapid transfer of money and growing amounts of it. Gold, awkwardly cumbersome, could not meet the needs. Abstractions, such as coins, paper currency, and letters of credit, were invented. Yet, gold—unlike money or letters of credit—is an eternal element. It is permanent, and the struggle for its possession is one of mankind's oldest dramas.

The dramas center around whoever has physical possession.

The six city men seized control of Willie Douthit's gold *the moment* they moved it from the cave to their own hiding place. Once hidden, the gold was secure, it would remain secure only so long as it was hidden.

Doc Noss broke that security.

We do not know precisely when or how Noss found the gold, but we may put aside as fiction his description of underground streams, torture posts, and twenty-seven human skeletons.

We can deduce from the interviews of people who knew Noss that he found the gold first and he found Victorio Peak later, his wife, Ova, being the first victim of his deception.

A surprisingly informative picture of Noss's activities between 1936 and 1949 emerges from the interviews and from the files of the New Mexico state land office, the museum of New Mexico, the Department of the Army, the Secretary of the Treasury, the Securities and Exchange Commission, and the Secret Service.

From these, we find Noss discussing treasure with his lawyer in 1936; moving it in 1937 and pretending to find it in Victorio Peak; and selling it steadily until 1939, when his cave is sealed by a miscalculated blast of dynamite.

After the cave-in, Noss begins a consistent effort to reopen the peak. He needs money, and he commences frauds. He is eventually killed as a consequence of these frauds.

Ironically, when Noss has access to his gold he is afraid of it—afraid to sell it, afraid for it to be known he has it. He is afraid of ambushers, of the "city men," of the law.

It is only after he loses the gold that he becomes convinced it is his and becomes bold in his efforts to recapture it.

Doc Noss and his wife of two years, Ova, who was known as "Babe," arrived in Hot Springs in 1933. He was twenty-eight years old, she thirty-eight. They rented a house and he set up his shingle as a chiropodist, trying to eke out a living. His few patients, however, would pay him off almost entirely in barter—mostly vegetables and some game. He saw little cash.

But it was in Hot Springs that his life took its first direction toward the treasure.

We can assume that upon arrival in Hot Springs Doc was soon made aware of the tales of Willie Douthit and Buster Ward. Indeed, the pair had vanished only a few months previously. Doc learned more details, perhaps embellished, when he met and became friends with a leathery, sixty-three-year-old prospector and miner, Claude Fincher.

Fincher and Doc began making trips into the Organs. It was not serious treasure hunting. It was mostly an excuse to get outdoors, to smell mountain air, to shoot some quail, and to see some scenery. Later, Fincher and Doc would become mining partners in the San Andres.

In those days Hot Springs was not far removed from the memory of the Wild West. People were alive who knew—some who even rode with—Pat Garrett, Johnny Ringo, Clay Allison, and Billy the Kid. The men of southern New Mexico in the 1930s commonly wore guns. Doc was one of these; he was usually armed, dressed in black, tall, and lean; and, unusual in the taciturn West, he was outspoken, quarrelsome, and rather a braggart. Despite such bad manners, he nevertheless had charm and a group of friends, particularly younger men. He looked dashing. He was dangerous.

Deputy Sheriff Ernest Tafoya made a point of keeping an eye on Doc. Tafoya found him to be "quarrelsome and violent." Years later, at Doc's inquest, Deputy Tafoya would be asked:

"Was his reputation good or bad?"

"I wouldn't say it was too good," replied Tafoya.

"Good or bad?"

"Bad."

For Doc, pickings apparently proved too lean in Hot Springs, and 1934 found him in Wellington, Texas, near Oklahoma in the Texas panhandle. There he was arrested and charged with practicing medicine without a license.

He skipped bond and crossed back into New Mexico. There, at Roswell, he had his quarrel with the waitress, "while armed." On July 15, 1935, he was sentenced to serve six months in the state prison. He was paroled, however, before the year ended.

Promptly upon leaving the prison, Doc surfaced again in Hot Springs. He arrived carrying a treasure map.

His old mining partner, Claude Fincher, was not in Hot Springs that year. With the apparent purpose of replacing Fincher, Doc sought out the friendship of Tony Carriago, a young Mexican prospector familiar with the Caballos. He told Carriago a vague story of having acquired the map in Mexico. He needed Carriago's help because of the Mexican's knowledge of the Caballos.

Carriago was persuaded to quit his job. The two men began searching the Caballos.

In the mountains, they encountered several times an ancient, gray-bearded prospector named Parr and his son Roscoe. The Parrs were searching not only for the Douthit cave, which the elder Parr believed to be a mere cache, but also for a cave holding an even larger amount of the alleged La Rue treasure. At this stage of the story, Parr and his son have no role other than their chance encounters with Noss and Carriago. The Parrs come onstage for speaking roles later.

And now, from late 1935 through the first half of 1936, we lose Doc's tracks. He drops from view. When he surfaces again it is in Gallup, New Mexico.

Gallup. So far as we know this is the only time in Noss's life in which he is connected with northwestern New Mexico. Perhaps he was continuing his treasure hunting. Gallup is a bare thirty miles due north of the ruins of Hawikuh, the ancient city of "Cibola," which attracted Fray Marcos, which caused the death of Estavanico, and which was besieged by Coronado. But it is an unlikely place to hunt for gold. There has never been any evidence of it there, and, with the notable exception of Fray Marcos' imagination, there have never been any persistent legends of treasure in the region.

Top: The only known published photograph of Fred Drolte, the one-armed pilot who played a central role in the creation of the project Operation Goldfinder. The photograph shows Drolte on March 16, 1966, being led to the U. S. District Court in El Paso, Texas, where he was arraigned for gun running.

Bottom: Joe Newman, representative of the Mescalero Apache gold claim, stands at the boundary of White Sands and points out the arroyo he secretly used to visit Victorio Peak (photographs 9, 10).

This Indian mural, an irreplaceable historical art object, was dynamited by gold
hunters who illegally entered Hembrillo Basin with the help of Army personnel in the
1970s. Standing in the crater created by the dynamite is John Snow, security guard
and chief range rider for the Army at White Sands (photographs 11, 12).

The Army did an enormous amount of mining and road building at Victorio Peak, from 1961 onward. The extent of the Army mining did not become known until Operation Goldfinder in 1977 (photographs 13, 14).

In 1977, radar surveys by Stanford Research Institute corroborated Doc Noss's story of a huge cavern and a shaft sealed by cave-in. "In view of the . . . so-far-proven consistency of the original Noss story a further exploration of the peak should be undertaken . . ." said SRI (photograph 15).

Another possibility is that Doc was trying to get as far away as he could from Hot Springs while at the same time remaining close enough to get there in a single night's trip.

If this were the case, then Gallup fitted the bill. In Gallup, unlike Albuquerque or Santa Fe, Noss was unlikely to have a chance encounter with someone from Hot Springs. The two towns are in different marketing areas. Geographically, however, Gallup and Hot Springs are only 265 highway miles apart. Even on 1936 roads, that represented less than an eight-hour drive.

By August 1936, Doc had opened a chiropodist shop in Gallup. He remained in Gallup for more than a year, and it was in Gallup that he sold his first gold bars.

Doc became acquainted with Melvin Rueckhaus, an attorney in Gallup who would become a loyal friend, perhaps the best friend Doc ever had. Rueckhaus, the young, educated professional, found pleasure in the company of the rough-and-tumble Oklahoman.

"We had," Rueckhaus would say later, "several items in common. Both of us had an interest in mining and prospecting; I was a lawyer and he was a fellow that had a propensity for all kinds of troubles."[1]

Noss had set up his shop in the rear of a clothing store owned by Otis Swinford. His wife Babe was not in evidence. Rueckhaus did not know her then. Presumably she and Doc were living apart.

Doc and Rueckhaus were frequent fishing partners, and on one trip, in the fall of 1936, Doc mentioned for the first time that he had found gold.

"We went on a fishing trip to Blue Water together," said Rueckhaus, "and he showed me a fluorspar[2] property which both of us considered to be an attractive prospect. However, the property owner wanted money and the property itself would take considerable funds to put into commercial production.

"At that time he first told me of the gold that he had found which he felt might be used in exploiting the fluorspar property. At this time he described in generalities the find he had made

---

[1] Letter dated July 7, 1976, from Melvin Rueckhaus to Norman Scott, hereafter referred to as the "Rueckhaus letter."

[2] Fluorspar, or fluorite, is a valuable mineral used in industrial processes.

[that is, gold bars] and indicated that there were problems connected with it that he wasn't ready to go into with me."

Had Doc found gold bars by the fall of 1936? If so, it changes our perspective on the events of a year later, of November 7, 1937, when Ova Noss witnessed Doc climbing out of Victorio Peak with a gold bar.

If Doc found gold in 1936, then he only pretended to find it in 1937, and he set up his wife as part of the deception.

That hypothesis depends upon whether he did find gold in 1936, and what purpose would be served by pretending to make the find a year later in Victorio Peak.

There is no reason to believe that he did not find gold in 1936. He did not attempt to sell gold bars or stock to Rueckhaus or, as far as we know, anybody else. Doc may, of course, simply have been trying to inflate his importance in Rueckhaus' eyes, a strong possibility with a man like Noss.

Nevertheless, Doc was probably telling the truth. This is supported by the fact that within ten months Rueckhaus would personally inspect a bar of gold, witness its assay, and witness the bar's sale for twenty thousand dollars. All of this occurred months prior to the dramatic events of November 7, 1937.

Not only had Doc sold a bar prior to that date, but also apparently he knew of the existence of the caves in Victorio Peak prior to that date.

In an interview given nearly forty years later, rancher Mart Gilmore said that Doc came to him in late 1935 or early 1936 asking to be shown some caves.[3] It was Gilmore who introduced Doc Noss to Victorio Peak.

Gilmore, born in Tularosa in 1903, was a regular visitor to Hembrillo from the age of two weeks, when he and his mother were first taken through Hembrillo Canyon in a covered wagon. He and his brother Moss became thoroughly familiar with the area during the years their father ranched there until 1927.

"My dad, Watt [sic] Gilmore, bought the old Fleck place in 1890, and Hembrillo is part of it," said Mart Gilmore.

"We used that basin for cattle in the winter. . . . When we lived at Engle about 1936, I took my mother to Hot Springs for

[3] El Paso *Herald Post,* Aug. 30, 1973.

Doc Noss to treat her feet. He asked me if I knew of any caves in the San Andres Mountains and asked me to show him, and I did. I didn't think much of it at the time, but I knew there wasn't any gold there."

Gilmore described the appearance of the canyon and the basin when he rode them as a cowboy. He said there was an old fort at the north end of the canyon where he and his brother used to pick up shells.

"I never saw an oxcart or stagecoach, as someone has supposed to have found there, and we rode every foot of that country. My grandfather, Jack Cravens, found about 75 weapons—burned—in the trees about 2½ miles south of the old fort, back before the turn of the century.

"We leased Hembrillo from a mining man named Westfall who had a lead mine there around 1900. It had one spring on the north side of Victorio Peak—and we never called it that; we knew it as Hembrillo Peak.

"My dad and I and my brother and most of the cowboys have been in the cave, and there is nothing in there and never was. The old lead mine on the peak goes straight down. Westfall had to quit mining it when water got in it. My dad bought the main spring from Westfall in 1902. There were several springs around there, and an old rock house north of the main spring there.

"There was no gold" in Hembrillo Basin, Hembrillo Canyon, or Victorio Peak. "No swords, no nothing."

Gilmore is vague about when Noss questioned him about caves. It was "about 1936" when Noss was a chiropodist in Hot Springs. Noss was a chiropodist in Hot Springs in late 1935 through the first half of 1936. Following that he moved to Gallup and returned to Hot Springs only a few weeks prior to his "discovery" of the caves in late 1937.

It appears then that Doc had gold and had explored the caves prior to November 7. Furthermore, rancher Gilmore said the caves contained "no gold . . . no swords, no nothing." Yet from November 7 onward, many people see, and some even assist, Doc Noss to take gold from the caves. A logical possibility is that Doc found gold elsewhere, moved it to the caves, then set up a sensational discovery for his wife to witness.

Why the work? Why the theatrics?
Let us consider the following:

Hypothesis 1: Doc finds a new treasure—not the Douthit lode, but another. He files a mining claim, or a treasure claim, and banks his profits. In New Mexico, mining claims and treasure claims can be filed on land owned by others, so long as it does not conflict with existing subsurface claims. In that instance, Doc would have no reason to "salt" Victorio Peak.

Hypothesis 2: Doc finds a treasure on a site already claimed, or contained within the boundaries of a military reservation. Doc would then have sufficient reason to salt Victorio Peak.

Hypothesis 3: Doc finds the Douthit gold. He now has the best reason of all to shift it to Victorio Peak. If Noss finds the cache of the city men and does not move it, they would kill him to keep their secret. However, if he manages to move it without detection, the power shifts to his hands. He not only has their gold, but he also has the power of their destruction.

The city men would be as handicapped as Willie Douthit in protecting their interests. They would have violated the Gold Reserve Act of 1933 in possessing the gold; and, since Doc would know it to be Douthit's lode, they conceivably would be liable as accomplices to whatever implication Douthit had with the Organ Mountain corpse.

By moving the gold, Doc possessed it. The city men could not proceed against him directly, using the law, without seriously jeopardizing their own freedom.

They could, of course, proceed against him covertly, recovering their property in a murderous and secret raid. But all concerned attest that Doc was prepared against just such an attack—prepared to the point of paranoia.

X

Pearl Harbor

In their tent, under blankets, Doc told Ova, "I've found something and I don't know what . . . a cave big enough for a freight train . . . a mummified skeleton with red hair grinning up at me . . . Wells Fargo chests, church relics, guns, swords, jewels . . . and enough silver and gold to load sixty to eighty mules. . . ."

From the information of the previous chapter, we may infer that in the winter of 1935–36 Doc Noss found treasure either on a military site or at the cache of the city men. In the same winter he persuaded rancher Gilmore to show him "caves in the San Andres," and Gilmore showed him Victorio Peak. In the meantime, Doc had moved to Gallup and become friends with young Rueckhaus. In the late summer of 1937 Doc probably begins caching gold at Victorio Peak, and possibly at other sites. Years later, Harvey Snow, the Tularosa postman, will be told by cowboys that they saw Doc move loaded mules *from* Hardscrabble, twenty miles north of Victorio, *to* Hembrillo, and leaving Hembrillo with the mule packs empty. This could be the period, about 1937, when Doc was salting Victorio Peak.

The lawyer Rueckhaus remembers distinctly that it was midsummer of 1937—late June or early July—when Doc Noss sold his first gold bar.

Both men had moved from Gallup to Albuquerque.

"Shortly thereafter," said Rueckhaus, "without previous notice to me, Doc had gone to Jack Levick at the Reliable Pawn Shop in the 200 block of South First Street, in an effort to sell Levick some gold.

"Levick was reluctant to deal with Doc, and Doc gave him my name as reference. Levick called me, and I was curious enough to go down to the pawn shop and witness the transaction.

"Levick's gold license as a pawnbroker allowed him to only purchase a limited amount, so it was suggested that Levick buy a small piece. Doc took a hunting knife out of his pocket and cut a slice off the bar. Levick used a karat tester and said that it was almost pure gold, something in the neighborhood of twenty-three karats.

"In the course of the discussion, Levick mentioned that another customer of his, one Mark Levy, had a friend, a priest, who would buy larger amounts of the material. At a later time the priest arrived, and Levick retested the balance of the bar, and the bargain was made that the priest would pay to Noss the sum of twenty thousand dollars.

"The priest delivered the money and took the balance of the bar. Noss went on a spree of drinking and spending and was depleted of all of his cash within a period of about three weeks."[1]

After sobering up from his three-week binge, Noss seemed unconcerned about the twenty thousand dollars he had squandered. There was plenty more.

In the same summer of 1937, said Rueckhaus, "Doc took me to his hotel room in the 100 block of Central, which was later to be known as the King's Hotel, and showed me six more bars, [which were] in appearance similar to the one I witnessed the test upon.

"They were blackish in color, about 2½ inches wide, 2 inches deep, and about 8 inches long. They were extremely heavy by weight.

"Doc told me that he had five more bars but that the other five were so marked that they might get him in trouble. Later he told me that the trouble he anticipated was from Wells Fargo, whom he stated claimed title to the bars. Noss implied that the marked bars were the loot of a robbery of a Wells Fargo coach many years before . . ."

In the late summer of 1937, still a few months prior to climbing Victorio Peak with Ova, Doc tells Rueckhaus of his plans to salt an old mine, and is possibly referring to Victorio.

[1] Rueckhaus letter.

"Noss told me," said Rueckhaus, "of a plan that he had in mind to dispose of the marked bars by salting an old mine in either the Caballos or the San Andres. His efforts in this direction were somewhat hindered by his constant complaints of people following him. . . ."

Rueckhaus also described another profitable sale by Noss. "Later in 1937 or 1938, Noss told me of selling another bar to one Ruby Horowitz of Nogales, Mexico. We talked several times about his efforts to collect the money due on the bar. This time the price was sixteen thousand dollars, and in the collection efforts it was settled for eight thousand dollars."

During this period, Noss told Rueckhaus that he was negotiating to sell a gold chalice to the priest who had bought the gold bar.

A few weeks later, said Rueckhaus, "I heard that the Bernalillo County sheriff's department had picked up a gold chalice. Noss asked me to check it for him and gave me a general description. I examined the gold chalice at the sheriff's office and found it to somewhat answer the description given me by Noss. However, Noss then said he didn't want to pursue the matter any further."[2]

Noss, as we shall see, will continue to avoid contact with law-enforcement agencies.

It was November 7, 1937, when Doc and Ova visited Hembrillo Basin as part of a hunting party. They had moved to Hot Springs only a few weeks previously. While Ova had sheltered herself from the rain, Doc had gone down into Victorio Peak to a depth of 187 feet. "He then proceeded along a passage to a 'room' 2,700 feet long. The room, he claimed, contained a large treasure: several stacks of Spanish-made gold bars; chest of jewelry; Spanish armor; swords; crowns; the statues of several saints; and Wells Fargo chests. Noss also reported that there were twenty-seven skeletons in the room, and later he brought one out to prove it. He said that from inside the room he had discovered an old entrance to it, blocked with logs and earth, that faced the southwest."[3]

[2] Ibid.
[3] Report by the division of research, Museum of New Mexico (Santa Fe, 1963). The report was authored by archaeologist Chester Johnson of the museum staff and represents the result of museum excavations at Victorio

Noss never succeeded in finding the southwestern entrance from the surface. He later stated, however, that at 2 P.M. the sun shone down into the long cave through a small hole at the top of Victorio Peak.

Noss then hired a local boy from Rincon, José Serafin Sedillo, to help him in the difficult task of bringing out the gold.

Serafin would later claim that he had found the cave years before he met Doc Noss.

As a boy, he had worked for the Henderson ranch, tending goats as they pastured in Hembrillo Basin. One afternoon, the goats were clambering over the peak, and one disappeared in a crevice. Serafin investigates and finds a hole. It is dark and mysterious. The boy is afraid to go in, but, fearing Señor Henderson more than the cave, he follows the goat.

It is a big hole and goes down a long way. I think about stories my father has told me. He has said many times a story his father told him. That there is a big rich mine in these mountains. My father and my grandfather both looked many years for this mine. When I told my father I had found a big hole on the peak he said he was too old to climb down the hole with me. He tells me to say nothing. When I become a grown man I will come back to the hole and explore.[4]

When Serafin did return, he found Doc Noss already at the scene. Noss hired him.

Serafin "made only one trip to the treasure room with Doc Noss, but on this occasion he carried out two gold bars in his pockets. Since part of the climb had to be made with the aid of ropes, no heavy load could be brought out at one time. When he reached the surface, Serafin said he was going to keep the bars he brought out. Doc said he wasn't and would kill him first. Doc

Peak in 1963. Johnson's information was derived from on-site examinations at Victorio Peak, state land office records, interviews with Gaddis mining personnel, interviews with Ova Noss, and interviews with other persons connected to the Noss treasure find. Hereafter, the report will be referred to as the "Museum Report."

[4] Henry James, *The Curse of the San Andres.* Serafin told James the story in 1952 while Serafin was prospecting at Victorio Peak. The above quote from Serafin is paraphrased to the extent that I have eliminated Mr. James's dialectical treatment of Serafin's speech.

pulled a gun. Serafin left and did not return until after the cave was blasted shut."[5]

After Serafin left, Noss hired Joe Andreg, a boy who had been working on the ranch of his uncle south of Victorio Peak. Joe Andreg was only sixteen when he worked for Noss, and he told a Museum of New Mexico interviewer in 1963 that "he cared nothing for the gold," but he "greatly admired the man, Noss, and his fancy, black western outfit." Andreg stated further that he went down only a little way in the passage, where he waited for Noss until the latter should return. He said that Noss took empty bags down and, when he came back up, the bags were filled with heavy objects.

Andreg said he saw no gold while at the cave. Later, at the nearby camp outside the cave, he helped to saw a bar into quarters. This bar was soft and could be bent, said Andreg.

At a meeting at the camp, a man unknown to Andreg asked Noss if he had "the crown." Noss replied, "Right here," and slid a satchel across the table to him.

The man opened the satchel, then a sack inside the satchel, and looked. He seemed satisfied as he closed the sack and returned the satchel to Noss.

Andreg never saw the crown, he said, since it was not his business to see it.

The last time he was with Noss, said Andreg, Noss asked him to accompany him to El Paso.

As they were leaving the Hembrillo Basin by the eastern pass, Noss stopped by a bank of earth along the road and pulled a heavy box into the truckbed. Noss was "broke" and borrowed money from Andreg to buy gasoline to reach El Paso. Upon arrival, Andreg went immediately to his hotel room, and when he next saw Noss, the latter was carrying a large amount in travelers' checks, and the box was gone. Then they went to Las Cruces, where Noss went on another binge and lost the checks.

Andreg said that seventeen hundred dollars of the checks were later found and returned to Noss by the Las Cruces police.

By the end of 1937, according to several witnesses, Doc Noss had taken from the cave eighty-eight bars of bullion, a crown, a

---

[5] "Museum Report," p. 3; interview with Ella Foster, former wife of José Serafin Sedillo.

naked female statue "originally thought to be gold but now known to be brass, a quantity of jewelry a hinge from a Wells Fargo chest, a tin box of papers, a sword, and possibly other items."[6]

According to Ova Noss, in the early months of 1938 she walked into a cafe at Clovis, in eastern New Mexico far from the San Andres, one evening and found Doc seated at a table. There was a beer in front of him and on the floor a tin box of yellowed letters and other papers.

"These were papers," said Ova, "that Doc had found in the cave. He had nearly finished reading through them when I came in. He told me to sit down. When he had finished, we went out behind the cafe. Doc put the papers in a trash can and set them afire. He said they would have allowed descendants of the people mentioned in the letters to claim the gold. Later he told me that the odd thing about those letters was that none of them was dated later than 1880."

In this period, Doc began to make inquiries in regard to staking a legal claim to Victorio Peak. He drove to Albuquerque and went to see his friend, the lawyer Rueckhaus.

"Noss had me accompany him to Santa Fe to try to get some sort of a state lease on a piece of mining property."

While in Santa Fe, Noss became "acquainted with a person in the state land office by the name of Mrs. Cave. In connection with this endeavor," said Rueckhaus, "Noss took me to the peak, which you described to me as being the location of his find. Later, in conversations with Mrs. Cave, she told me that she had gone down and had entered the cave but that the powdered guano was so strong that it made her violently ill and she had to leave. She described a gold sword that Doc Noss had shown her during one of these visits. . . .

"During this same time frame, I suggested to Noss that he cease to be afraid of Wells Fargo and others and try sending some of the gold to the mint where it could be identified and turned into cash. He agreed to attempt this and told me that he had. I did not handle any part of this endeavor."

In the spring of 1938, the Secret Service received a letter concerning Doc Noss. That letter, dated April 11, was the first entry

6 "Museum Report," p. 3.

in what would prove to be a voluminous dossier, and from the public's standpoint a frustrating one.

The letter was from Hubert Heath, the editor of the Hot Springs *Herald*. It is repeated here in its entirety:

> *United States Treasury*   *April 11, 1938*
> *Intelligence Division*
> *Washington, D.C.*
>
>   *Gentlemen, on the evening of March 29 word was relayed into my office that 1,000 ounces of gold had been seized here by the Chief of New Mexico State Police.*
>
>   *The report is that the gold was taken from a fellow known locally as Dr. M. E. Noss. In the past he has used the name Knoss, Nos and Moss, the latter said to be the name under which he did time in Oklahoma, according to a man who knew him there.*
>
>   *Noss or Moss is affiliated with the———[7] His wife told a friend of mine that the gold came from buried treasure he had found. A young man who is employed in our shop went to Juarez and interviewed revolutionaries and was told the gold was sent across to pay for guns for the Gold Shirts, or at least for revolutionaries.*
>
>   *Very little has leaked out about the matter but he was told that great quantities of guns were being crossed, greater than at any time in the past and that they were being paid for in gold dust.*
>
>   *[Signature of Hubert Heath deleted by Secret Service]*

The *Herald* letter marks the beginning of the Secret Service's role in the White Sands story. For the next thirty-eight years it will play an important part in the story. And over that thirty-eight-year period we will see it perform with steady incompetence and deceit. We will find a lot of liars, a lot of whitewashing, in the Secret Service.

Its first investigation of the Noss story consisted only of an interview with the chief of the state police. The Secret Service did not talk with the *Herald* editor, the original informant.

The Albuquerque agent reported that the chief denied any sei-

[7] The deletion on the public record was made by a Secret Service censor.

zure of gold. The chief said he knew Noss, and that "Noss was suspected of handling narcotics." As for the *Herald* editor's complaint, the agent attributed it to local politics in Hot Springs. Noss was a friend of the incumbent mayor, said the agent, and complaints against Noss by the *Herald* were "evidently prompted by the fact that the Hot Springs *Herald* very much opposes the administration of Mayor————[8] in Hot Springs, New Mexico. . . . Accordingly, the file is being retained and the case closed in this District."[9]

Noss meanwhile was happily unaware of the federal scrutiny. His gold activities continued. One report comes from B. D. Lampros, who was working then with Noss at Victorio Peak. Later, Lampros would give an affidavit on his experiences, saying:

"In 1939 I took one rough bullion bar on my round to Douglas, Arizona, and had Holly & Holly to assay it and it was found . . . to run over five thousand dollars. . . ." He also describes posing for a picture in which he and "Colonel Willard E. Holt of Lordsburg, New Mexico, were holding one-half bar each as the bar was sawed in two. The bar was regular Wells-Fargo gold bar. I saw quite a few other bars of the same type."[10]

In March 1939, the chief of city police in Albuquerque, Pat O'Grady, wrote the state prison inquiring about Noss's record. O'Grady explained that Noss and Ova were living in Albuquerque at the time. "A short time ago, Noss disappeared mysteriously [from Albuquerque]. We heard rumors of him having a gold mine in Sierra County, also that he has uncovered some hidden treasure."

In July 1939, Ova Noss wrote the United States Mint in Washington, D.C., inquiring about the law of gold ownership. The undated letter was done on the letterhead of the Hotel Baker in Deming, New Mexico, and stated in part:

"We are residents of New Mexico and find lots of stories about buried treasures such as bars of gold and silver—also Mexican money. Now we have in our possession an old map that tells of a rick of gold bars and Mexican money. Now we are searching in

[8] Deleted by the Secret Service.
[9] Secret Service report on Dr. M. E. Noss dated May 31, 1938.
[10] Lampros affidavit dated October 19, 1952, and submitted to the Department of the Army as part of an application by Ova Noss to dig at White Sands. Hereafter referred to as "1952 Noss Application."

the vicinity of where it is liable to be—and what I want to know is what [is] the United States law on a find of that kind? Can we take it to the United States Mint and sell it or will the government claim the treasure?"

The director of the Mint, Leland Howard, bounced the letter over to the Secret Service for action. That agency replied stiffly that if Mrs. Noss had gold she had to report it, advising "that the finding of any gold in this country by any person should be reported promptly to the Treasury Department, together with a statement as to the circumstances under which it was found, in order that specific instructions may be given."

The Secret Service failed to tell the Nosses that they would be reimbursed for surrendering gold. Had the letter been *calculated* to frighten off gold owners, it could not have been more successful. Alarmed, the Nosses backed off.

Meanwhile, the Secret Service dispatched an Albuquerque agent to interview Mrs. Noss. Over a three-month period he failed to find her. He ended up by leaving a message with Ova's son Harold Beckwith, to give to his mother.

By August of 1939, Noss had hired an engineer named S. E. Montgomery to facilitate the removal of gold by enlarging the cave opening.

Noss kept the shaft entrance covered and locked. No one was permitted to enter unless Noss was present. No work or exploration was done during his absence.

Noss wanted to dynamite a larger entrance, and Montgomery was fearful of a cave-in. The engineer was not permitted to make an examination sufficient to prove or disprove his theory.

Montgomery believed there already had been a partial cave-in at the shaft. Noss himself earlier in the year had almost caved in the shaft when dynamiting a boulder. Several tons of the loose talus covered and filled the passageway. This the workers were removing.

As the work proceeded, Noss was hitting the bottle and was impatient to the point of abusiveness. After some argument, Montgomery agreed to load the charge in the amount Noss wanted. But Montgomery said the shaft should be timbered first, and timber couldn't be laid until the workers had finished removing the talus.

Noss would have none of this slow work. He was in a hurry.

Benny Samaniego was one of the workers present the day of the blast.[11] He had sneaked into the treasure room. He lit his candle and caught his breath.

"I saw stacks of gold bars, skeletons, armor, old guns, and statues. The skeletons were tied, kneeling to posts, as if they were prisoners left to die."[12] Samaniego scratched his initials in one gold bar.

Then, from the direction of the shaft, he heard Doc and Montgomery arguing. "They were arguing about the place to put the charge. Montgomery said Doc didn't know what he was doing. I thought they were going to blow it right then, and I was about to yell and let them know I was there. But they decided to wait until the next morning."

Samaniego sneaked out of the treasure room without, he said, taking any bars.

The following morning, Montgomery and Noss set off their blast. To the horror of everyone present, the shaft caved in. The treasure was sealed off.

Noss went off on a bender. He may have assumed that the loss was temporary. He certainly had no idea that he would never again see the treasure room.

Meanwhile, Benny Samaniego bought a house.

In 1963, Samaniego told a Museum of New Mexico interviewer that he had not removed gold from the cave. But there is a suggestion of modest wealth to be found in the Don Ana county clerk's office, record book 96, p. 337.

It states that on September 8, 1939, within two weeks of the cave-in, the previously penniless Samaniego bought a house and lot at 937 South Melendres Street in Las Cruces. The value was $16,000.

When Samaniego died in 1970, his estate had total assets of $73,672.[13]

The cave was sealed in August 1939. For the next two years, Doc Noss attempted to raise money for reopening it. He aban-

---

[11] His family claims descent from Lope de Samaniego, a knight who was second in command to Coronado and killed early in the Cibola expedition.

[12] "Museum Report," p. 4, interview with Samaniego.

[13] Last Will and Testament of Marguerite G. Samaniego, filed in District Court, Dona Ana County, May 13, 1974.

doned the effort in the spring of 1942, when the Army closed all roads leading to the San Andres and sealed off an Air Corps bombing range.

During the balance of 1939 and throughout 1940, Doc kept a minimum of three men working daily to clear the shaft. Doc was paying for the operation out of his own funds, and in early 1941 he ran out of money.

In February of 1941 Noss was at the peak to close down operations when the prospectors Parr arrived at Hembrillo Basin. They were as surprised to see Noss as he was to see them. Since 1935, they had been searching the Caballos for the La Rue treasure. Now they had obtained new information, which led them directly to Victorio Peak. They had no idea of finding Noss there.

"My father and I believed we had located the area of the La Rue mining settlement . . . when we got there we found Dr. Milton E. Noss, who was exercising squatter's rights over the area we desired to investigate."[14]

The Parrs and Noss exchanged greetings. The Parrs then left.

Noss was worried. He had failed to file a mining claim on the Hembrillo site. The Parrs conceivably could file, stake claim, and bar him from his own cache. That night he went into Hatch, New Mexico, to find the Parrs. He carried with him a bar of gold.

Noss offered a share of the treasure to the Parrs in return for their financial assistance in clearing the shaft. He presented the gold bar as evidence of good faith. The Parrs dug out bits of the bar in several places and the next day had it assayed. It assayed at better than 60 per cent, or fourteen-karat gold. A written contract was signed, and the Parrs and Noss became partners.

In June 1941, the partnership was expanded to include, among others, Claude Fincher. Fincher was the old prospector who had guided Noss in his original tour of the Caballos in 1935.

A decision was made to create a company. To obtain stockholders, Doc brought prospective clients to Victorio Peak, where he would show them gold. Some of these stockholders later signed affidavits as to what they saw. Among them was C. D. Patterson, mayor of Portales, New Mexico:

"I believe I know Doc M. E. Noss as well as any person living

---

[14] Letter to President Richard Nixon from Doyle Elliott, attorney for Roscoe Parr (July 6, 1973).

today. I have had several heart-to-heart talks with Doc, and I felt he spoke the truth as to his find in Victoria Peak. He was more or less seeking help as to how to get the dirt out of the cave, which was hindering his getting into it. He offered to show me what he had taken out before it caved in, and said he had taken out almost one hundred bars. When I visited Hembrillo Basin in Socorro County in July 1941, there were a lot of people out there helping him, and I took the opportunity to walk with Doc.

"We didn't walk for long when he went over to a bush and pulled out a bar from underneath. I was amazed at what I saw. This bar was sawed *into,* and looked like gold all the way through: beautiful, yellow, golden. It was longer than it was wide, and was about three inches thick, and weighed around twenty pounds. The whole bar would have been about forty pounds."[15]

Another prospective client to visit the peak was Don Breech. He stated:

"At the time I visited Doc Noss at El Humbrillo [sic] Basin in July 1941, I owned the Ford Agency in Portales, New Mexico. I, along with Mr. C. D. Patterson and Mr. E. F. Foreman, all of Portales, visited there for several days. Doc needed a truck of some kind to haul supplies out to the camp, so I loaned him one of mine. He kept it for several months, but after his help left the truck was returned. I had so much confidence in the possibilities of Doc's adventure that I contributed several dollars to be used for the purchase of groceries.

"While I was visiting there, I heard Doc tell the story of his find, which later he confirmed with proof. I was shown a sword, napkin ring, old money, and a fan, which I had never seen one like before. And last, but not least, several bars, which looked like gold. Some of these had holes drilled in them and showed the same contents as deep as the drill hole went."[16]

Other visitors in July 1941 would tell the same story.

A more skeptical witness was Deputy Eppie Montoya. In 1949, at Doc's murder inquest, Montoya would be among those to testify that Doc had a bad and dangerous reputation. In 1941, however, Montoya worked on Victorio Peak as a deputized guard.

In a 1952 affidavit, he stated:

[15] "1952 Noss Application."
[16] Ibid.

"I, Eppie Montoya . . . was employed by B. D. Lampros. In May 1941, I was deputized by the U.S. marshal to go to work for Doc M. E. Noss at the Hembrillo Basin in the San Andres Mountains. . . . During my stay there, I had the opportunity and privilege to see some bars supposed to be gold and yellow in color, which he had taken out of this cave in Victory [sic] peak, one of which I held and examined.

"It was brick form, larger at bottom than at top, slanting up on all sides, and weighing approximately twenty to twenty-five pounds. Doc told me that there were a great many more in the cave that had caved in. Dirt and rubbish had to be removed to get to the room. At the time that I saw this bar there were four or five people present."

Noss, Ova, the Parrs, Lampros, Claude Fincher, Willard Holt, Don Breech, C. D. Patterson, and twelve other persons formed the Cheyenne Mining Company. The purpose of the company was to finance excavation of Victorio Peak *and* elsewhere in the San Andres.

Shares were allotted on a percentage basis and, for an unknown reason, the largest shareholder was a Los Angeles law firm, Robinson & Barman. Ova Noss was the second-largest shareholder, with 18.78 per cent; Doc was third, with 10.67 per cent. The Nosses held the voting proxy of the Los Angeles firm, and as a consequence, control of the corporation. A fourth major shareholder was one Jack Bruton, a Rincon rancher who had beaten both Noss and the Parrs to the state land office and filed claim on Victorio Peak. For his waiver of claim, he received 5.22 per cent.

The first meeting of the Cheyenne Mining Company was held in Las Cruces on June 14, 1941. The second and final meeting was held six months later, at the Rock House in Hembrillo Basin.

All of the shareholders, and many members of their families, were present. Some children were there, playing on the slopes of Victorio Peak. A secretary kept minutes.

It was a festive occasion. People had brought beer, and picnic lunches were laid out. The talk was exciting. Noss announced that the company had obtained the co-operation of the state and the federal governments in excavating the claim. The property itself had recently been turned over by the state to the U. S. Department of the Interior. Noss said soldiers from Fort Bliss had in the past month made nine trips to the site to inspect it.

A deputy sheriff had been hired to stay on the grounds permanently. His duty was to run off trespassers. The excavation had aroused great excitement in the nearby communities, and would-be gold poachers had become a problem.

Willard Holt said he had received an offer to sell a claim for $22,000, which he refused.

Mart Gilmore was present. He was the cowboy who had originally shown Noss the cave six years earlier. Gilmore knew it to be empty then but was so convinced it now contained treasure that he had put up five hundred dollars for a portion of a share.

It was announced that a major offer had been made by an El Paso group. The Texans had offered $35,000 for all the rights to talcum deposits at Hembrillo; they had offered $32,000 for silver rights, and $150,000 for all contents of the treasure cave. It was voted to refuse the offers.

Then Noss stood up. According to the recorded minutes, "he discussed the progress of the work, saying they were down 264 feet in the shaft, and that it was going to take some money, time, and labor yet, but he thought that within two weeks they could have it ready to open if the materials were provided."

Two weeks, and the cave would be opened.

The date of the meeting was Thursday, December 4, 1941. Three days later—on Sunday, December 7—the Japanese bombed Pearl Harbor. The San Andres were sealed shut. Hembrillo Basin became part of a bombing range. Three and a half years later its slopes would be lit white by the glare from the explosion at Trinity Site.

XI

Murdered

Hatch, March, 5, 1949 (AP)—M. E. Noss of Hatch was shot to death today. Charlie Ryan surrendered voluntarily to Dona Ana County officers. Sheriff Apodaca received the following account of the shooting.

Noss, labor foreman for Ryan's mining company, and Ryan engaged in a loud argument over their business. About 3 P.M. Noss hit Ryan and knocked him into a window. Then Noss ran out of the house toward a pickup truck owned by the mining company. Just as he reached the truck, two shots were fired. One of the bullets hit Noss in the right cheek just below the eye and lodged at the base of the brain. . . . No charges have been filed.

An inquest was held five days later, on March 10. One of Doc's workers, Willard Lee Blake, was an eyewitness to the shooting. He testified, "Doc came running out of the house and run toward the pickup. Charlie Ryan come in behind him, and so did Charlie's wife. Charlie fired one shot and said, 'You come out from under there, you sonofabitch!' His wife spoke up and said, 'Shoot the sonofabitch!' And he fired that last shot."

Blake's testimony contrasted sharply with Ryan's reputation. Ryan was a hard-working, churchgoing man who owned a machine shop in Alice, Texas, and did oil rig repairs. He had met Noss only four months earlier when Noss was penniless and out of work. Ryan had taken pity on him, played the Good Samaritan. Now, ironically, he had ended up killing him. Ryan was ordered held for trial.

Following the inquest, but prior to the trial, with suspicious timing the Secret Service leaked a report on Noss to the Albuquerque *Journal*. It was not their own report but an FBI memo stating that Doc Noss "has operated in the confidence racket, old gold brick and gold mine swindle and buried treasure swindle."

In addition to that leak, the Secret Service senior agent in Albuquerque, James Hirst, also commented for attribution upon Noss's "fake bars" in several interviews. At the trial, Ryan would be acquitted on the basis of justifiable homicide; Noss's reputation was shredded, and the general tone of the press and the trial itself was that Ryan had done the community a service.

The truth, of course, was that Doc was never accused of any gold swindle until after the 1939 cave-in.

Noss did some swindling in the later years. But the proceeds went to his excavation project. He died with $2.16 in his pockets. In the meantime, he had spent ten years of hard work trying to dig out the gold. He raised capital, hired engineers and labor, even carved out a road to the peak capable of handling a pickup truck.

Trial testimony, and a statement given at the time by Ova Noss to the state land office, suggested that Doc was shot for betraying Ryan.

Doc was killed, said Ova, because Ryan had caught him removing and hiding fifty-one bars of gold. Would this account for Mrs. Ryan's uncharacteristic language? Her fury? *"Shoot the sonofabitch!"*

Ova claimed that Doc had removed the fifty-one bars and hidden them on different slopes of the Hembrillo Basin. He wouldn't tell her where the caches were, and he wouldn't cash in the gold. Why? Ova explained, "Doc was not educated, and he always had a feeling that if his discovery was known, he would be punished by law-enforcement officers or physically harmed by robbers. . . . He didn't know how to get any cash on this gold since there was a ban on owning it . . . he tried to dispose of it through the black market but had little success."

The last ten years of Noss's life is a tale of irony. In 1939 he had established his claim to Victorio Peak so thoroughly that he was able to hire workers and an engineer. On nearly the very eve of commencing his large-scale recovery, the shaft was closed by a cave-in.

By the end of 1941, after two years of excavation financed largely out of his own pocket, he was again within reach of the treasure. He announced that he was within two weeks of a break-through. Three days later Pearl Harbor was attacked, and he was denied access to the site.

He returned after the war, overcame numerous difficulties, and in 1949 seemed once again to be at the very brim of success. He was instead shot dead, gunned by a man who had befriended him and had been his benefactor.

Noss's Cheyenne Mining Company became inactive following the December 4, 1941 meeting at Hembrillo Basin. Doc, in the meantime, somehow avoided the draft, and 1942 found him installed in Del Rio, Texas, as a radio personality.

Melvin Rueckhaus saw him infrequently during this period, but he remembers:

"Doc became involved with a Dr. Brinkley in Del Rio, Texas, and had set up a business of selling astrology readings through radio advertising. He stated that he was doing well under the stage name of The Great Karrem.

"During the war years, I heard from Noss irregularly and only saw him on one occasion. We went to Chihuahua, Mexico, together to handle a problem of some confiscated property of his acquired during the Del Rio days."

Rueckhaus said that Noss also was selling gold stock during the period:

"Doc had taken to selling fractional interests in his treasure venture and also appeared to be doing well selling certificates."

Rueckhaus saw little of Noss thereafter. "I did not see him again until sometime shortly before his death, when he discussed with me the problems he was anticipating with Roscoe Parr and Charlie Ryan."[1]

On July 16, 1945, the bomb was exploded at Trinity Site. Two months later, on August 14, Japan surrendered, and World War II was ended.

Six weeks later, to retain possession of the San Andres, the Army filed condemnation proceedings against the Department of the Interior, which held title to some of the White Sands range; the state of New Mexico; and private owners. The Army asked for "exclusive use of the land for military purposes" and, rather

[1] Rueckhaus letter.

curiously, also asked for "full and exclusive possession of all oil, gas, and other mineral interests."[2]

The condemnation was immediately contested by the Department of the Interior and the state. The litigation would go on for eight years. In the meantime, Doc Noss had access to Hembrillo Basin, and he wasted no time getting there.

In January 1946, he applied for mining claims at Hembrillo. The state land office replied that the claim would be given a routine examination the following month.

On February 4, consulting engineer Gordon Herkenhoff arrived on assignment for the land office.

Herkenhoff and Noss spent several weeks together, and Herkenhoff found Noss "a likable man who knew and was known by everybody. . . .

"He was a swarthy, well-built man about five feet, ten inches tall and very friendly. Doc had surrounded himself with a group of real tough cowpuncher types from the Clovis area who were all armed and looked quite mean." Noss wore a gun, too, a fact that saved Herkenhoff's life.

Herkenhoff and Doc were climbing on Victorio Peak to obtain rock samples when the engineer put his hand under a small ledge.

"Hold still!" Noss shouted, pulling his .45-caliber pistol and shooting, dead in the head, a large rattlesnake that had been coiled under the ledge, inches from Herkenhoff's hand.

Twice the engineer descended with Noss about 186 feet deep into the peak before being stopped by a land fill. Noss explained that it was the consequence of the 1939 dynamite blast. He described the huge cave within and the find of vast amounts of gold bars.

The engineer studied the exterior of the peak and concluded that at one time horizontal entrances had existed at the base and had subsequently been filled in, either naturally or by man's design. If there were a huge cave within, it would have been a perfect sanctuary and storehouse for Apaches, bandits, Padre La Rue, or whomever.

"It might have been possible," said Herkenhoff later, "to take things into the sides of the split, from the base of the peak, con-

2 Letter to Congressman George Mahon of Texas from Major General H. A. Gerhardt (Jan. 2, 1962).

ceal them there, and then fill in the split from the other side"—that is, from within the cave—and climb out through the top of the peak.

Noss told Herkenhoff, as he had told other persons earlier, that he had found blocked passages that appeared to have led horizontally from the treasure caves.

The pair went to Hatch and had some supper and some beers while Noss elaborated on the treasure. He described gold bars, Spanish treasure chests, church relics, jeweled crowns, and precious stones. He estimated the value at $22 million.

Out of the corner of his eye, Noss studied his man. Gordon Herkenhoff's eyes were glazed.

Well, New Mexico does call itself "The Land of Enchantment." The title is deserved. It has ghosts. It is a land walked by the ghosts of elephant-hunting men and by the creative geniuses of the Clovis and Sandia points. It is the land of conquistadors and the priests of Cibola working their mystical electric light machines. It is the land of Billy the Kid, Pat Garrett, the Gray Ghost, and Lozen, White Painted Woman. New Mexico does enchant. Its very terrain makes magic. New Mexico fires the imagination with fantasies.

Despite the gaudy dimensions of Noss's story, Gordon Herkenhoff, consulting engineer for the land office, bought it—but not *quite* all the way.

He told Doc he would be interested in investing some money to help excavate the caverns—but only if Doc would produce some gold bars for his inspection.

Noss said he would produce them. The men separated.

Noss had no gold bars. He had sold the last of them in 1939 to finance the excavation of the cave-in. And when his money ran out in 1941, he began taking in partners. These included the stockholders of the Cheyenne Mining Company, the men Doc took on walks and then casually produced before their eyes two "gold" bars.

Over the years, Doc had kept the bars in his suitcase. They were old friends. Warriors all. Survivors of successful combat. Now, at Herkenhoff's request, Doc brought them once again onto the field.

Herkenhoff remembers that a few days later "Doc called me

and asked me to meet him at the Park Hotel in Socorro. He told me he would then show me some gold bars." He arrived at the hotel to find Doc in the cafe and holding two golden bars wrapped in newspaper.

"He was gloriously drunk," Herkenhoff recalls. "He kept talking all the time and occasionally handed me the bars so I could feel their weight, each time taking them back and wrapping them again."

One of Noss's cowboy friends came by at that moment. Noss staggered to his feet, held a brief conversation, and the two wobbled off to find a bottle of whiskey. Noss left the wrapped bars behind.

Herkenhoff grabbed one of them and rushed over to the New Mexico School of Mines, located about six blocks away. He took the bar to the laboratory and tested it himself.

"It was pure brass. Mostly copper with some iron," said Herkenhoff.[3]

Herkenhoff lost all further interest in the investment. Noss's claim nevertheless was granted to Doc; his wife, Ova; and their old partner Willard Holt.

At Hembrillo, Noss felt he needed only a few weeks to break through to the cave. To do this, he hired a single laborer, José Serafin Sedillo, who moved with his wife to the old rock house near the peak.

Doc kept a close eye on Serafin. It was Serafin who as a young worker in 1937 had attempted to carry off two gold bars and was stopped by Doc at gunpoint.

The work went slowly, starting up whenever Doc obtained financing, dying when he didn't. In 1946–47, the Sierra Talc Company bulldozed a road from the Jornada to Hembrillo Basin and across it to service a talcum mine two miles northeast of Hembrillo. This road, abandoned by the company a year later, gave Doc for the first time easy automobile access to Victorio Peak. The road passed within a few hundred yards of the base of the peak. He had, however, no money to exploit this convenience.

In 1947, Doc—on some journey or other—fell in love with a pretty Arkansas girl, Violet Boles. In an Arkansas court, the

[3] The Gordon Herkenhoff quotes are obtained from an interview contained in the story "Test Proves Doc Noss's Metal Bars Pure Brass" by Howard Bryan in the El Paso *Herald Post* (Aug. 23, 1973).

forty-four-year-old Doc forthwith divorced the fifty-four-year-old Ova, claiming desertion: "Her whereabouts are unknown to me." He then married young Violet.

They went to Corpus Christi, Texas, to find work in the oil fields. In the meantime, Doc had carelessly allowed his claim to lapse.

As for Ova, she was by no means missing, "her whereabouts" unknown. She was on the scene at Hembrillo, and when she learned of the divorce, she raced to Santa Fe to renew the mining claims in her name, not Doc's.

In response to the claim, land office engineer Donn Clippinger examined Victorio Peak. He concluded that there was a high probability of gold content.[4] But he also found a second treasure, which everyone had overlooked:

"Another valuable deposit within the cavern has previously been overlooked by Mrs. Noss and her associates. In Mrs. Noss's discussion of the examination of the caves by Dr. Noss, she made several references to a 'bad odor' in many of the chambers and stated that Dr. Noss in some parts of the caverns was 'obliged to walk knee deep through a fluffy brown material which had an unpleasant aroma, irritated his eyes and nostrils and caused his legs to itch and break out.' This 'brown fluffy material' with its irritating effect on the skin is undoubtedly dry guano deposited there by bats that formerly inhabited the caverns, as at Carlsbad and elsewhere. This dry, light condition shows that the bat guano has not been leached by water and retains its high content of phosphate, nitrate and potash. Bat guano, the most valuable of all animal fertilizers, will find a ready market. . . ."[5]

Noss went broke in Corpus Christi, and with no prospects, he drifted around looking for work. He roamed far, ending up penniless in the deep southern tip of Texas, in Jim Wells County, a place as barren and desolate as any in New Mexico—except, and it is a big *except*—it is a place of oil wells. There, in the county seat of Alice, he found work. He was *given* work. It was given to him by his killer, Charlie Ryan. Later, Ryan described the circumstances.

[4] See Chapter I.
[5] "Report on Proposed Noss State Mineral Leases" (Santa Fe: state land office, Oct. 13, 1948).

"We needed a truck driver to go out and pick up tools. We ran an ad in the paper. Doc answered the ad. The head bookkeeper, who is in charge when I am out of town, interviewed him and hired him. I was in Houston at the time. When I came back that is when I first met Doc."

Doc's duties were to go out into the oil field and pick up work that needed to be repaired by Ryan's shop. Doc worked on a commission basis, without salary, and was supposed to furnish his own truck. He told Ryan's bookkeeper he had a good one.

"When I came back," said Ryan, "he had an old 1935 model truck. I looked at the truck and I told him I couldn't let the truck go out with my name on it because that would be a bad reflection on my business. He said he had the money and would trade it in on a new one. He fooled around about it, and I told him he better get started. He had been working for the Gulf Refining Company in Corpus Christi, he said, and they hadn't paid him. He claimed they owed him $761, so he asked me would I loan him the money to buy a truck. I told him that I couldn't do that.

"I didn't know the man, but he was a good talker, and he got me over to the First National Bank at Bishop, Texas, and talked to Mr. Buck, the vice president. Mr. Buck told me Noss and a fellow named Smith, who owned the Gulf Refining Company, had been there, and Smith told him he owed him some money."

Ryan questioned Buck about Noss. "I don't know much about him," Buck replied. "He has done a little business here, he and this fellow Smith."

Ryan, against his better judgment, decided to help Noss buy a truck. They went to Corpus Christi, where "we was looking at some trucks at a car lot when a police car drove up and the police asked who was driving Noss's car and he walked over and put him under arrest. I was in my car and I got out and asked the officer what the trouble was. He said, 'Swindling, by worthless checks.' They took him to the jail, and I went on to the jail. The attorney said they had five checks people had turned in for collection. So I related the conversation I had had with Mr. Buck. I told him I believed this fellow was sincere when he wrote the checks, that Smith admitted owing him [more than] $700."

In reply, Ryan was informed by the county attorney that Noss still owed $400 on the 1935 truck.

At this stage, with Noss's lies unfolding like the leaves of an artichoke, even an averagely naïve man would begin to suspect that something was wrong. He would back off. The most charitable saint would back away. A St. Francis or a Mary Magdalene would tuck up their skirts and take off. But not Charlie Ryan.

Confused by Doc's deceptions, compromised by his own Protestant charity, Ryan believed Doc to be innocent until proved thricefold to be otherwise. Ryan backed up his trust with money.

Ryan first agreed to make good on Doc's outstanding bad checks. Toward this account, he paid the county attorney $150 and secured Doc's freedom.

Then they went to a car lot together and Ryan cosigned a note allowing Doc to buy a $1,760 pickup truck. With Doc driving the truck, they returned to Alice.

"When we got home," said Ryan, "there was a long-distance call there for him. It was a man in McAllen who had cashed two $50 checks for him; he was going to turn him over to the law if he didn't pay them.

"I hated to see him go to jail. I called Mr. Buck and told him to let the checks come on through for $100, that I would pay it. That took care of that deal. That evening, I guess it was after quitting time, I was in the office and he came in. I told him, 'The next time a person calls up here about a hot check, you don't have a job. I don't put up with people drinking, giving hot checks or not paying their bills.'

"He said that would be the last one.

"He was complimenting me on what a nice business I had, said everyone spoke well of me. He said, 'This is a mint if I ever saw it.'

"I said, 'Yes, it's like owning a gold mine.' "[6]

That probably was just what Doc was thinking. The next morning, early, he told Ryan he had a gold mine himself.

Ryan was interested. Indeed, he felt that he now had so much invested in Noss that he couldn't afford to give up on him.

"I loaned him money to pay his rent, and then I come to find out his clothes was in hock—and his wife's wedding ring was in hock!

[6] Transcript of Noss inquest, Dona Ana County (Mar. 10, 1949). Hereafter referred to as "Noss Inquest."

"I gave him money to get that out.

"He got a wire one night from someone that wanted to see him, said that his wife was getting a claim on this mine. He come in there the next morning and told me he just had to go to Santa Fe. He told me this story while I was in bed with a sprained ankle."

Ryan got out of bed, and with Doc driving Ryan's flashy yellow Buick convertible, they speeded to Santa Fe. They were too late. They were informed at the land office that the permits had already been granted to Ova. A land office attorney asked Ryan to stay over until the next day so that he could be interviewed concerning any false claims Noss might have made.

"I told him I didn't want to get into anything that would get me into trouble. . . .

"Doc said I was the only honest man he had known. . . . He wanted me to go in on the mine with him. I told him, 'I will give you a check for $1,000 and this pickup truck. You go up there and then when you get your money you can pay me back.' He said no, the people up here [in New Mexico] would beat him out of it because he couldn't get anything in his name. I never had a chance to talk to anybody about the man, other than about those bad checks.

"My wife, she said, 'Charlie, if there ever was a person I feel sorry for it's this fellow Noss.'

"He would come in my house and talk until ten or eleven o'clock about this thing. I let him live in a garage apartment I had. I was trying to help the man. They was charging him $17.50 a week in a tourist court. He wasn't using it, and I told him to move in. I furnished him lights, gas and water, it wasn't costing him anything. I was trying to help the man like I have others, and I have had to ask for help myself."[7]

By January, Noss had convinced Ryan to accompany him to New Mexico and begin working the claim. Noss also brought in Roscoe Parr as a partner.

Upon arrival at Hembrillo Basin, they found they had been preceded by a considerable population. Encamped at the site, on her claims, was the still-angry Ova. Staying with her was her son by an earlier marriage, Marvin Beckwith. Encamped at the rock

[7] "Noss Inquest."

house was José Serafin Sedillo and his wife, who were now working for Ova.

The Noss-Ryan party made camp on the slopes adjoining Ova's claim. In the meantime, both men rented houses in Hatch and installed their wives.

After a few days of what we can imagine to have been a very delicate negotiation, Doc soothed Ova and worked out a compromise whereby she would allow them to excavate the claim in return for an equal share.

The entire month of February was given to digging. Six workers were engaged, including Serafin and the foreman, Willard Blake.

Noss was at the mine every day. He was always near a gun. When he wasn't wearing it on his hip in a holster he had it in the glove compartment of his pickup. Once, to demonstrate his skill to Blake, he shot a moving rabbit at sixty yards.

The men worked at the peak only during daylight hours, arriving at the site about eight-thirty in the morning and leaving about four-thirty in the afternoon so that they did not have to travel the Jornada in the dark. Noss, however, made many trips to the peak at night, trips that triggered many suspicious conjectures on the part of the workers and on the part of his partners.

It seems that at this time he found not the main cave but one of his old caches, a rich one containing 110 gold bars weighing about 40 pounds apiece.

On the last day of February, Noss informed Charlie Ryan of his find. Noss stated, however, that the cache contained 51 bars, not 110. On behalf of the partnership, said Noss, he would take the 51 bars to Mexico, where he had already made arrangements to sell them on the black market.

No, said Ryan, they would not engage in anything illegitimate.

Noss argued, bullied, persuaded—went through, we can imagine, his entire con-man repertoire, but to no avail. Ryan insisted that the bars be brought to his house. They would then inform the proper authorities. The evening ended with Noss promising to deliver 51 bars the coming Saturday, March 5.

Friday morning, March 4, was the beginning of the last 36 hours of Doc Noss's life.

He arrived at Victorio Peak in the predawn hours. It was

deserted. Serafin and his wife had been sent the day before to Santa Fe to file some papers. The workers would not arrive for about three hours.

Doc began moving the rocks concealing his cache. Then he sat and waited. He was waiting for the sound of an airplane engine.

He had no intention, it seems, of delivering 51 bars to Ryan for surrender to the government. He had his old fears in that regard —fears, it will be seen later, that may not have been misplaced.

His intention was to fly the entire cache of 110 bars—slightly more than two tons of gold—out of Hembrillo and down to a pre-arranged site, where it would be transferred to his Mexican contacts.

How much of this plan, if any, was disclosed to Ova or her son Marvin Beckwith is unknown. But Marvin arrived at the Hot Springs airport that morning to meet a pilot, Curtis Noble, and his plane, which Marvin had chartered.

The plane, a Piper Cub, was capable of flying the 40 miles to Hembrillo, landing, loading, and returning to Hot Springs within 40 minutes, 45 minutes at maximum. It could lift slightly less than a ton of the bars per trip, and there was ample time to make three round trips before the workers arrived.

Meanwhile, Doc waited at Hembrillo. The hours passed, and by the time the workers arrived the plane still had not appeared.

We do not know Doc's reaction. We may assume it was one of jaw-grinding rage.

At 3 P.M., a plane appeared over the basin. It waggled its wings and flew low. There had been mechanical problems, and this had caused a delay. Beckwith was preparing to drop a note to Doc, apparently to inform him of the problem when, suddenly, the Piper hit an air pocket. It nose-dived into the ground with a fearsome crash and explosion.

Doc ran to the wreckage already afire and pulled out Beckwith and Noble. He loaded them into the pickup and sped to the hospital. There, Noble died that evening. Beckwith survived.

At seven o'clock that night, Noss gassed his truck at a Hatch service station and phoned an old friend, Leo O'Connell, in Santa Fe. Leo, a construction worker, was on a highway project, and his wife took the call.

Noss was very excited, she said, and told her, "The renegades are after me—I got to have Leo help me move that gold—they are going to fly it to Mexico tomorrow morning."[8]

She explained that her husband was away. Doc, in the meantime, looked for other help. Leo O'Connell would eventually arrive in Hatch too late, several hours after Doc was dead.

It seems that Doc had not given up his plans to shift the gold to Mexico. But the unexpected crash of the Piper Cub had presented an immediate problem. He had to get it out of the Victorio Peak area before daylight because sometime during the day Civil Aeronautics Board inspectors would arrive to inspect the plane-crash wreckage. Noss didn't want them to see gold.

Staying in Hatch that night was Tony Jolley, a cowboy and rodeo rider who knew Doc slightly. Jolley is tall, taciturn, slenderly graceful, and a gentleman. He is the type of Westerner exemplified by Gary Cooper in the movies.

In 1977, he told his story slowly, deliberately. The words were chosen with care. Jolley, by 1977 a wealthy Idaho rancher, did not want to be thought mad nor to be disbelieved.

"I was passing through Hatch," he said, "on the rodeo circuit. I had met Noss two or three times in Albuquerque, and at a filling station in Hatch that night I ran into him again. He asked where I was staying, and I told him a particular cabin. Later that night he came to the cabin. I was surprised. He said he needed help, needed it badly. Would I help him? I said sure.

"I got into his truck and we started into the Jornada. On the way he said that the next day he was to deliver bars to his partner Ryan, who had a plane waiting to fly it to Mexico. But Doc feared that Ryan planned to sell the gold, then scoot with the money. So Noss planned to sneak it away from Victorio and hide it elsewhere.

"It took us maybe three hours to cross the Jornada, and we must have got to the basin about ten-thirty or eleven o'clock at night. We went first not to Victorio but to the Dick Hill ranch, some ten miles this [the western] side of the basin. There Doc uncovered some brush and began handing me bars and told me to load them in the truck. There were twenty of them, each about forty pounds and dark-colored. I asked him what they was. He said: 'Gold.'

"We took them *into* Hembrillo Basin. *Into it,* you understand? We planted them at that cliff which has the Indian drawings about two miles north of Victorio.[9] After that we came to the peak and drove about halfway up on that road Doc had made. He had ninety more bars laying there under bushes. We took these and drove back to the cliffsite and buried sixty of them. The other thirty we carried out of Hembrillo. We went west some twelve miles and planted them in a meadow near a ranch house."

Based on Jolley's description, a reconstruction of the events of that night show that Doc had a cache at the Dick Hill ranch that he transferred back to the area of the mural. He and Jolley then went on to Victorio and removed ninety bars lying under bushes there. Sixty of these were also taken to the mural area. Another thirty were planted in Jornada north of the Dick Hill ranch. There were no bars left in the area of the Piper Cub wreckage.

They finished at about dawn. Noss, without sleep, dropped Jolley at his rooming house and went to the house of Charlie Ryan. He would try to excuse his failure to deliver the bars by the accident of the plane crash.

A light was burning in the kitchen of Ryan's house. Noss knocked. Ryan opened the door. He had a gun in his hand, and he *ordered* Doc to come in. He entered the kitchen and saw that Ryan, his wife, and two of the workers were present. Ryan was furious. He accused Doc of trying to sneak the gold away. Doc denied it. Ryan accused. It went back and forth for six hours. Finally, at two o'clock in the afternoon, Noss launched a punch at Ryan and broke through the kitchen door, racing for his pickup truck. He was going for his pistol.

Ryan knew it. Fearing for his life, and for the life of his wife, he grabbed up the gun. He shot at the running Doc. He hit him.

Doc fell beside the truck. He could not get into it without exposing himself to further fire. Instead, he scrambled under it, a wounded creature seeking his burrow.

The threat at that instance was ended. Noss had no means to kill Charlie Ryan.

[9] This is a reference to the area of the Indian mural.

Ryan took a step closer and aimed. What happened next is described in the eyewitness testimony of Willard Lee Blake:

Blake: Ryan said, "You come out from under there, you sonofabitch!" His wife spoke up and said, "Shoot the sonofabitch!" And he fired that last shot.[10]

Noss was dead. The main legendary figure of the region for fourteen years was gone. The dark, tall, dangerous man admired by Joe Andreg was no more.

We will hear no wails of remorse at his passing. Indeed, only three people seemed to have cared: his young widow, Violet; his friend Rueckhaus, who would assist in the funeral arrangements; and Roscoe Parr. Parr, a partner but never a particularly close friend, was the one who put up the money for Doc's funeral.

In the weeks following his death there was a scrambling among the potential heirs for his fortune. There was also a rather curious effort by federal agents to blacken Doc's reputation prior to the trial of Charlie Ryan for murder.

On April 7, a month after the shooting, the New Mexico Bureau of Mines confirmed to the Associated Press that there was a large cavern in the area of Hembrillo Basin. On the same date, James Hirst, Secret Service agent in Albuquerque, confirmed that he had taken a bar from Noss's effects. He had "visually" examined it and found it to be "defintely not gold." This was probably one of the two bars that Noss had shown to the Cheyenne Mining investors in 1941 and to Herkenhoff in 1946.

On April 9, Mrs. Merle Horzmann was declared administrator of Noss's estate and filed an inventory of his personal property. Mrs. Horzmann was the secretary of the inactive but still existing Cheyenne Mining Company.

According to Mrs. Horzmann's sworn statement, much of the personal property was in the possession of the Secret Service. The Secret Service would later deny any seizure of Noss's effects, including seizure of the metal bar that their own agent, James Hirst, told the Associated Press he had in his possession.

The mystery persists to the present day. Despite rigorous inquiry from congressmen, lawyers, and journalists, as late as 1977 the Service continued to state, in bureaucratic language, that

[10] "Noss Inquest."

Noss's bars were not seized. Agent Hirst was not available for questioning. He died in 1970.

The sworn inventory of Noss's possessions states:

Personal Property: One metal steamer trunk containing papers, documents, books, maps and other miscellaneous items, held by United States Government, Secret Service Agent, Albuquerque, for investigation;

One metal strong box, containing documents, papers and miscellaneous items, also being held by U. S. Government, S. S. Agent, Albuquerque. Supposedly three bars of metal, believed to be held by the U. S. Mint in Denver, said to have been receipted for to M. E. Noss, in the amount of approximately $90,000 since 1940 or 1941.

Supposedly, 51 bars of metal of undetermined value or content believed to be in possession of the U. S. Government Agent in Albuquerque, together with the above named three bars, pending investigation and evaluation.

Two bars of metal, presumably in Del Rio, Texas, at the Val Verde National Bank, value not known. . . .

Due to circumstances of cause of death of said Milton Ernest Noss, the United States Government, Secret Service Department, seized all personal property for the purpose of investigation, James W. Hirst, Albuquerque, New Mexico, agent.

On the following day, an FBI report was gratuitously given to the Associated Press. The report cited Noss's arrest and prison history and listed several vague and unattributed accusations against him as a gold swindler.

In May, Charlie Ryan went on trial. Melvin Rueckhaus was appointed special prosecutor.

Ryan testified that he had invested $27,000 in Noss's excavation. He testified that Noss was supposed to deliver 51 bars on March 5 and had failed to do so. But the death, he claimed, was a matter of self-defense. The jury decided not that it was self-defense, but that the homicide was justified. Ryan was acquitted.

Two months later, Agent Hirst was transferred to Los Angeles.

In the meantime, 110 bars of some sort of metal lay buried near the mural and in a meadow on the Jornada. More bars, presumably thousands of them, lay hidden in the depths of Hembrillo Basin.

XII

The
Cookie Jar

. . . the principals entered the cave and emerged at approximately 11 A.M. at which time they stated they could go no further; that the final forty-foot passage to the alleged gold was blocked by large boulders that could not be removed by hand or shoveled away. . . .

The principals all stated that although they were blocked at this time, they would not give up their efforts to recover this alleged gold hoard. . . .

. . . This operation has been carried as a top secret project. . . .

—Communication dated March 10, 1961, from senior agent, Albuquerque, to chief of Secret Service, Washington

. . . Colonel Jaffe stated . . . there was definitely no mining or sub-surface work being done by Military Personnel. He stated that the Army was not in the mining business . . .

*—Legal memorandum filed with
the state land office (Dec. 2, 1961)*

In the months following Doc's death, hundreds of treasure hunters invaded the Jornada looking for the bars. Day and night they prowled and dug in the area between Hembrillo Basin and the

Dick Hill ranch. Only the more daring invaded the White Sands reservation, for that was patrolled. None entered the basin itself. It was protected by armed guards and searched only by Ova and the partners. As far as is known, none of the bars were found.

By 1950, the amateurs had given up. Only the more terminally afflicted remained. Among these were:

•*Dick Richardson,* one of Noss's many ex-partners. In July 1950, he wrote an "inventory" stating that Doc had cached 5,600 gold bars at Victorio Peak. An even larger cache—11,980 bars—was located in a covered mine within Geronimo Peak, a mile north of Victorio. In the 1970s, Richardson's "inventory" would be made part of an official application for site entry.

•*Harvey Snow,* an oil field foreman who after years of working in Central America and the Far East settled at Tularosa in 1950. A native Texan, he became the most expert of the hunters at evading Army patrols. For the next twenty-five years Harvey Snow entered and left White Sands virtually at will. He conducted interviews and made his own explorations. Independent of Richardson, his conclusions were similar: Victorio was a cache; the larger cache was at Geronimo. Furthermore, he believed that the huge cave described by Doc Noss truly existed and was the original source of the gold. It was located, however, not in Hembrillo but in Hard Scrabble Mountain, fifteen miles to the north.

In 1951, the Army removed Ova and Serafin from Hembrillo. In terms of physical activity, the next seven years were dormant. Ova and the partners attempted repeatedly to obtain access to the site, through litigation and through the intervention of congressmen and the state land office. All the efforts failed.

In 1956, the Army's hold was further strengthened when it obtained an extension until 1970 of its leasing-by-condemnation.

In November 1958, another major find was claimed. This time it was by military men.

The finder was Leonard Fiege, a jet fighter pilot at Holloman Air Force Base, which is part of the White Sands military complex.

Captain Fiege described his find in an affidavit that was subsequently corroborated by an Army-administered polygraph test. In the sworn statement, Fiege said:

"In November of 1958, I, Captain Leonard V. Fiege, along with Ken Prather, Milt Wessel, and Tom Berchett were hunting on the White Sands Missile Range.

"We had separated and I went down a canyon by myself. I saw this small hill and open caves. I climbed the hill and found a small cave that was fairly well hidden.

"I was to meet the other boys at the base of this hill, but they had not yet arrived, so I decided to go into this cave and look around.

"The cave was large enough to stand up in until I came to a small opening about thirty inches around. I shined my light into this hole and it looked like it opened up into a large shaft on the other end.

"I climbed through on my stomach and entered the large part of the shaft again. This shaft led into the main cavern. The dust was too thick and the air was foul and hard to breathe, so I sat down on what I thought was a dust-covered pile of rocks.

"I was a little sick and was going to get out of there as soon as I stopped coughing. The dust fell off the pile I was sitting on, and I started inspecting this area.

"The pile of so-called rocks was not rocks but smelted gold in bars about the size of a house brick. I realized what I had found and grew a little excited. I wanted to explore more of this area, but in my excitement I moved too fast and the dust started flying.

"With my flashlight I saw three piles of this gold all lined up and another pile off to the left that was partly covered by the wall of the cave that had fallen in.

"In trying to shine the light around the cave to see if there were any more stacks of this around, I found it almost impossible to see farther than about fifteen to twenty feet because of the dust in the air. It was like headlights on a car shining into dense fog.

"I started getting sick again and made my way back out. I was determined to take the boys back in with me when they came to this hill.

"I was dirty and sick when they arrived. I told them what I had found and we agreed to go in and look at it again. We got as far as the small opening together, but Ken and Milt were too large to get through, so Tom Berchett and I went in alone.

"The dust we walked in after we slid through this tunnel was about five inches deep, and we stirred a lot of it up. We went in to these four piles of gold and confirmed the findings.

"We got sick again and moved back out of the tunnel and talked to the other boys. Tom and I [had] handled the gold and thought about taking some of it out, but decided against it because we were not familiar with laws that governed the claiming of this gold.

"We might lose it all if we took it out now, so we decided to go back in and cover this area up so nobody could find it. If we could find it, so could other people with enough nerve to go back in this cave past the small passageway.

"We decided the best thing to do was to close off this entrance with rocks and dirt. Tom and I re-entered this cave and worked for about 3½ hours off and on between going back to the small tunnel to get air. We caved in the roof and walls to make it look like the tunnel came to a dead end. Then we left.

"We have tried many ways to get into this area to remove the gold but we were always refused by the commander of the range.

"I believe that with gas masks or some air-filtering device we can work in the shaft with little difficulty. If more of this shaft has caved in, other than what we did ourselves, it will take time to remove this so we can get a clear passage to recover this gold.

"The piles were about four feet across the bottom and about three feet high in triangle-shaped piles. Because of the dust, we did not see any more, but we believe that there is much more in the area.

"We could string light and run air into the cave once we get through our cave-in. This would give us a chance to find any other treasure that might be in there with the four piles that we saw."[1]

Following their discovery, Fiege and his companions sought legal advice from Lieutenant Colonel Sigmund Gasiewicz, judge advocate at Holloman Air Force Base. He advised them to file a claim to the gold as treasure-trove and took an active interest in assisting them.

[1] Fiege's affidavit is contained in a memorandum from the chief, U. S. Secret Service, to the director of the Mint (July 3, 1961).

They made repeated entry requests, but the White Sands command—presided over by Army Brigadier General J. G. Shinkle—refused. Finally, in June 1961, a formal request was made by Fiege's commander, Air Force Major General Monte Canterbury.

The Air Force fared only slightly better than had the civilian applicants. Shinkle gave a polite but definite "No" to Canterbury, his superior in rank.

"Dear Monte," he wrote, ". . . I have caused a thorough inquiry into the applicable law to be made by my Staff Judge Advocate. His efforts included research and consultations last week in the Office of the Judge Advocate General of the Army in Washington . . . [who] recommended that Army authorities should not permit any person to remove property of this character with which we are concerned from a military post. This is because the Department of the Army has no authority to recognize title or bestow rights upon private parties to treasure-trove found on a Federal reservation. . . .

"I must deny Captain Fiege permission to enter the area at this time for the purpose requested. . . ."

Shinkle was more helpful to the Air Force than he had been to the civilian applicants, however. He suggested a method by which his decision could be appealed: Apply to the Secretary of the Treasury and the Secretary of the Army.

Fiege and his group moved to do just that. On the day of Shinkle's rejection, a contract was signed between the treasure finders —Fiege, Berchett, Prather, and Wessel—and a legal team of Air Force lawyers—Gasiewicz, Ralph Garman, and Edward Rice.

The treasure finders would receive 20 per cent each of all proceeds, the legal team 5 per cent each. (The remaining 5 per cent was reserved for contingencies.)

The partnership called itself the "Seven Heirs."

Two weeks later, the Seven Heirs met in Washington with representatives of the Treasury Secretary, the Army, and the Secret Service. The meeting was presided over by the director of the Mint. At the conclusion, it was decided to excavate at Victorio Peak.

The Secret Service advised that the project proceed "as discreetly and cautiously as possible."[2] Nevertheless, within two

[2] Memo from the chief of the Secret Service to the director of the Mint. (July 3, 1961).

weeks, *Life* magazine contacted the Secret Service in Washington. A *Life* reporter wanted the name "of the captain who discovered the gold . . . in New Mexico."

"I told him," said the Secret Service inspector involved, "it was our policy not to discuss cases under investigation and gave him no other information. He [the reporter] said he knew how to get it anyway, and I suggested that he obtain it that way."

Belying the pretense of unconcern, the Secret Service conducted a short investigation to find the leak. It was unsuccessful.

Plans for the excavation moved ahead. On July 31, the senior agent at Albuquerque, Pat Boggs, was instructed to accompany "military personnel for the recovery of alleged gold in an abandoned mine in the Andres Mountains.

"The Bureau of the Mint suggests that if a recovery and seizure is made it might be stored at a new Federal Reserve Branch Bank in El Paso, Texas. Further, any assay of portions of the recovery can be arranged through either the Phelps-Dodge Refining Corporation or the American Smelting and Refining Company, both of which have proper facilities in El Paso. It is also suggested that if a recovery is made, that you communicate with this office telephonically."[3]

On the morning of August 5, Agent Boggs drove to White Sands headquarters. There he met Major General Shinkle, Captain Fiege, Berchett, Prather, and Wessel.

They rode to the basin in Jeeps. Accompanying them in a military police Jeep were four MPs. Ten more armed MPs followed in a weapons carrier, which also contained picks, pickaxes, shovels, and other light digging tools.

Fiege had been prevented from entering the site for nearly three years, from the time of his original application. Now he and his partners found themselves disoriented.

"The information supplied by the principals varied as to the distance they had to travel and how they had originally caved in the tunnel," Agent Boggs reported. "Captain Fiege was interrogated unknown to the other principals. However, he still maintained that he had seen and handled the bars. . . ."[4]

[3] Communication from chief, Secret Service, to senior agent, Albuquerque (July 31, 1961).
[4] Communication from senior agent, Albuquerque, to chief, Secret Service (Aug. 10, 1961).

For four days they attempted to find the cave. Finally, on August 8, Fiege said he had found the entrance.

Fiege went in and penetrated one hundred feet into the peak. There he encountered a cave-in and said he could proceed no farther. General Shinkle sent the smallest of the MPs in to confirm. It was true. The entrance was sealed.

Agent Boggs had become disgusted with the whole adventure. His reports, skeptical from the beginning, abruptly took on the tone of an agent preparing an arrest report. Fiege and his partners ceased to be called "principals." They became "subjects." And from a perusal of the reports, one infers that they were suspects in a criminal conspiracy. The nature of the conspiracy is not stated.

Boggs "suggested" to Fiege and Berchett "that they submit to a polygraph examination to authenticate their statements that they had acually seen and handled the bars of yellow metal. These are the only two subjects to enter the chamber. The subjects were somewhat evasive in their reply to this suggestion; however, they did state they would discuss it" with their attorneys.

The project was marked "pending" until results were obtained from the lie detector tests. In the meantime, Boggs assured Washington, "this operation has been carried as a top secret project; and the only persons at WSMR (White Sands Missile Range) who knew the nature of this mission were [censored].[5]

The test was given at White Sands on September 13, 1961. The results were positive. In November 1958, Fiege had entered the cave. He had seen and handled bars of yellow metal that he believed to be gold. He had seen four huge piles of such bars. Every aspect of his affidavit held up. Airman Berchett also submitted to the polygraph. He had identical results.

On September 20, the Secretary of the Army ordered the excavation of the tunnel. Hembrillo Basin, haven for Apaches and deer, changed its face, from the primeval and undisturbed to an Army beachhead. The area was cordoned off by MPs and patrolled by helicopters. Brought to the peak were bulldozers, trucks, weapons carriers, Jeeps, and mobile homes for barracks. Brought, too, were timbers to shore up the shaft, compressed air motors, digging machinery, and electrical generators. A road was carved up to the crest of the peak, and the bulldozers cut out a

[5] Ibid.

helicopter pad. The preparations took two weeks. Then the digging began.

The entire operation was illegal.

New Mexico law states unequivocally that no mining, treasure-troving, or mineral exploitation can occur in the state without authorization of the state land office. The law applies regardless of who owns, leases, or occupies the land in question.

Later, the Army would acknowledge the primacy of the state law in the matter. It is not known if the Army, or the Secret Service, knew of the illegality at the time they authorized the digging. The Secret Service in later years would refuse to release documents pertaining to the legality. They likewise refused to release papers showing how or why the project was marked "Top Secret," the highest security classification in the Secret Service.

It all went for naught.

On October 31 Ova Noss reported to the land office that military personnel with mining equipment were working her claim.

The man who received that complaint was Oscar Jordan.

Commencing that day, he set in motion the machinery of the state of New Mexico to settle once and for all whether there was gold at Hembrillo Basin. For the next fifteen years, he would be blocked at every turn—by the Army. He would be lied to and deceived—by the Army.

Oscar Jordan is a tall man, looking like a bank president. He is a model civil servant, efficient and competent and charming. He is trustworthy. And he has power. Yet the Army would deal with him as if he were a foreign agent until Oscar Jordan's legal powers were aroused. And the Secret Service would attempt to dazzle him with a blitz of fuzzification.[6]

After receiving Ova's call, Jordan dug out an old memo of his, dated three months earlier, that stated that he had been phoned by "Colonel Sigmund I. Gasiewicz, Staff Judge Advocate at Holloman, [who] said that some military person had found a large number of gold bars and that he had one in his possession." Gasiewicz

---

[6] "Fuzzification" is a word invented by James H. Boren of Washington, D.C., president of the International Association of Professional Bureaucrats. The word, says Boren, means "the presentation of a matter in terms that permit adjustive interpretation. Particularly useful when the fuzzifier does not know what he or she is talking about, or when the fuzzifier wants to enunciate a nonposition in the form of a position."

did not say he had seen the bar. He had asked Jordan if he could inspect the state's treasure files, including the Noss file. Jordan had agreed.

On the day of Ova's complaint, Jordan telephoned White Sands only to be told that Colonel Gasiewicz had been transferred to Washington and had been succeeded by a Colonel Thomason. Jordan phoned Thomason, and the parade of lies began.

Thomason advised Jordan that the Army did have a file on treasure-trove at White Sands. However, Jordan wrote in his file memo, "he was not familiar with it; he had heard rumblings that some people were doing some work. I told him that I was interested in getting ahold of the military agency responsible for permitting access to that part of the range where this gold was supposed to be located; that, although we claim no right to go upon that part of the range except to protect our interests, that the government had condemned the use of the surface and their right to take the minerals, but that the minerals were still state-owned. He advised me that White Sands, under General Shinkle, had custody of the range."

Jordan stated that he was dispatching a land office engineer to White Sands to "investigate as to whether or not unauthorized persons were mining this state property. If we found anyone mining on this state land, whether they be civilian or military, we intend to put a stop to it promptly."

The balloon had gone up.

On the following day, Ova and an attorney went to White Sands headquarters and demanded to see the general who was jumping her claim. She was referred to Colonel Jaffe, an Army lawyer and spokesman for the commanding general. Jaffe told her that he knew of no mining. She could not, however, inspect the area because "it was a closed area and was under the surveillance of the Army to exclude trespassers. The Army is making tests there and carrying on necessary activities on the Range related to the military mission."[7]

A few days later, Oscar Jordan's mining investigator was given the same denial by Colonel Jaffe. Jordan recorded it in a land office memo dated November 6:

"I called Colonel Jaffe and asked permission to make investi-

[7] Memo from Colonel Jaffe to Secretary of the Army (Dec. 6, 1961).

gation of the land down there in question to determine if any activity on the part of unauthorized persons had taken place. Colonel Jaffe was not very co-operative; however, he did agree to contact the commanding general and call me back. I advised Colonel Jaffe that if authority was not granted for a representative of this office to make an investigation we would attempt to secure a court order from the U. S. District Judge for authority to go up on the land.

"After informing Colonel Jaffe of our position, he attempted to assure me that there was no operation, that it was all a myth."

And again Jaffe told Ova's son Harold Beckwith the same story.

Said Beckwith, "When questioned about the discovery of certain treasures from the Noss mine and the work that was being done on the Noss mine, Colonel Jaffe denied both queries. He stated that he had been on the base a year and that nothing had been exposed in the like of a treasure or minerals on the White Sands base and that there was definitely no mining or subsurface work being done by military personnel. He stated that the Army was not in the mining business. He also denied any inquiries from the land office pertaining to the matter."[8]

On November 6, the very day that Jaffe told the land office it was all a "myth," Ova Noss walked into Oscar Jordan's office and slapped four affidavits on his desk. "Now let's see who's a liar," she said.

The affidavits were from Ray and Bob Bradley, H. L. Moreland, and R. B. Gray. On October 28 they had slipped up to the rim of Hembrillo and witnessed the Army's mining.

Ray Bradley saw "two light plants, one of which was running, several five-gallon GI gas cans, a walkie-talkie, mining wedges, and timber. I also saw light plant cables running into a cave."

Bob Bradley saw "military Jeep No. 0987057 . . . another military vehicle which is commonly called a weapons carrier. . . . The Jeep was a military police vehicle with a red light on the hood. It had a two-way radio set. . . ."

Moreland and Gray made similar observations.

Jordan passed on the new information to Colonel Jaffe.

The following day the Army halted its mining. The Secret

---

[8] Letter from Robert Martin, attorney for Ova Noss, to General John G. Shinkle (Dec. 2, 1961). The letter is contained in a state land office file on White Sands. Hereafter the file will be referred to as "land office file."

Service was not too happy. Its Albuquerque agent recorded sourly that Oscar Jordan "exhibited to me supposed affidavits by trespassers on the missile range alleging extensive mining operations by the Army. . . ."[9]

The Army pledged that no more mining would be done. But in December, Ova and her "trespassers" returned to Victorio Peak and found two military men still at work in the cave. They reported it. Jordan made his phone call. Finally, the mining ceased.

Jordan now offered to have the state of New Mexico supervise the mining. The Army could resume its excavations with a legal permit, and an inventory would be made of all treasure and artifacts removed. The Army refused. It would not submit itself to state authority.

Through the medium of memos, the two sides quarreled for the next several months. Then the Museum of New Mexico offered its good offices. *It* would supervise the mining, and the excavation could be done by a civilian firm. All the Army and state needed to do was grant permission.

Jordan and the land office agreed promptly. The Army continued to resist.

In the meantime, as New Mexico's two U.S. senators and a platoon of Texas and New Mexico congressmen intervened on behalf of the museum, there were three interesting side developments:

•In New York City, two men—Stanley Lane, 46, and Joseph Valle, 39, were arrested for possessing 2½ bars of gold bullion valued at $8,000 each. Both had Phoenix, Arizona, addresses, and the Justice Department said they were members "of a ring that has unlimited access to" a western gold mine. The pair had been arrested in a New York apartment. The bullion was in a suitcase. At the time of arrest, the U.S. attorney said, "The government regards this as a most serious offense, particularly in view of the very serious situation this country faces regarding its gold balance." In future years, neither the Justice Department nor the Secret Service would release *any* details on the case, save to say, "It was never prosecuted."

•On November 14, 1962, Mr. and Mrs. Roy Henderson went to see Oscar Jordan and asked to review the White Sands file. "After

[9] Memo from Albuquerque to Washington (Feb. 27, 1962).

reviewing the file, Mrs. Henderson left the office and asked Mr. Henderson to tell me some of the things he knew about the purported Noss treasure-trove. Mr. Henderson talked very little, however, he finally stated that he and his wife own a forty-acre tract adjoining . . . [Victorio Peak]. . . . He stated that he had been into the cave approximately eighty feet, which was as far as he could go because the crevice became too small. He stated that on the walls were markings, pictures, and drawings. He also stated that he had seen a large number of purported precious jewels, which were supposed to have been taken from the mine. Among other things shown Mr. Henderson by Doc Noss were some type of an old broom; a Spanish sword, which appeared to be of antique origin; and two bars, which were supposed to be gold. These bars had green corrosive spots on them."[10] They were, of course, Doc's old friends, the brass bars. Gold does not corrode.

•In the same month, Harvey Snow, the Tularosa man who had been exploring the San Andres since 1950, got sore feet. Snow had taken advantage of the attention focused on Hembrillo to penetrate the Organ Mountains behind White Sands headquarters. In the sheer ravines overlooking the headquarters he found a silver vein "so rich I could cut away strips with my knife. I packed seventy-two pounds of silver into my pack and carried it down the cliffs, down waterfalls, down the mountain. When I got home, I couldn't hardly walk for a month." A few months later, Harvey bought a couple of acres of residential lots.

The year 1962 closed with no decision on the museum's application. In July 1963, the Army finally succumbed to congressional pressure and allowed the museum sixty days to excavate and "determine positively whether or not there is treasure" at White Sands.

The Museum of New Mexico would not find treasure. It would not even find Fiege's cave. That cave would not be seen again until 1973, when Joe Newman and a young partner, Kevin Henry, entered the Fiege cave.

What is interesting about the museum's excavations is the Army's attempts to sabotage them and later to misrepresent the findings.

[10] Memo, land office file (Nov. 14, 1962).

XIII

Cape Work

The results of the exploration program conducted in 1963 proved the existence of a number of open caves within Victoria Peak similar to those described by the individual who claimed to have been in the caves and seen the artifacts and treasure.
>—Museum of New Mexico report (March 11, 1965), filed with commanding general, White Sands

. . . the [museum] geologist concluded in his report that no caverns such as claimed by Doc Noss were revealed. . . .
>—Commanding general, White Sands, in press briefing (Aug. 1973)

The Army granted the excavation permit only through intensive congressional pressure—particularly from Texas' powerful George Mahon, chairman of the House Appropriations Committee.

In the ensuing dig, the Army would treat the museum much as a matador treats a bull. The bull is encouraged to charge at something real only to encounter empty air. Government agencies call such tactics "cape work."

The Secretary of the Army granted the museum a thirty-day license, later extended to sixty days, to excavate at Victorio Peak, commencing on July 20, 1963. The actual digging was to be done by the Gaddis Mining Company of Denver, Colorado. The private firm, working on speculation, had agreed to shoulder all costs in exchange for 35 per cent of any profits realized. Gaddis would bring in thirty-five professional miners, drilling rigs, earth-moving equipment, and seismographic equipment. The company would spend one hundred thousand dollars, bearing costs even of liability insurance and the payroll of Army security guards.

The on-scene supervisor for Gaddis was Loren Smith, president of the company. The museum's on-scene man was archaeologist Chester Johnson, Jr. Ova Noss was technical adviser for the project. The Army, however, forbade her entry to the site.

There were numerous restrictions. The area in which exploration of any type was permitted was limited to the triangle within the three arroyos around Victorio Peak. No one was permitted outside the triangle. No cameras were permitted. And permission to camp within the basin was limited to seven men. All others in the project camped outside the missile range boundary.

The miners' first project was to drive a tunnel 218 feet into the peak in an effort to intersect the caved-in shaft reported by Fiege.

Censorship was imposed from the beginning of the operation, and it was a month before word leaked to the public. In an August 23 dispatch, the Associated Press reported "that a shaft had been sunk into a cave without any discoveries being made . . . the museum is involved in the exploration in accordance with the Federal Antiquities Act and under a permit from the state land office . . . there is a section in the agreement between the Army and the museum that no information would be released to the public unless mutually agreed upon by both parties. . . ."

This amounted to mutual censorship, and the Army would later exercise its option in the museum's final report.

As the digging continued, the miners found only a few pottery sherds, some broken metates of the prehistoric period, and some arrowheads dating from around the year A.D. 1, the time the nearby volcano had exploded.

By the end of the first month, while work was continuing on the 218-foot tunnel, "all other crevices, shafts, tunnels within the peak had been thoroughly explored and evaluated as to the amount of time and work it would take to enter the caverns through them. . . . Doc Noss's old passage was rejected as it was decided that there had been too much digging around and changing of its feature in the twenty-four years that it has been closed. To have used Noss's old passage would have meant that all the rubble and

Norman Scott, professional treasure hunter, leads a column of newsmen down from Victorio Peak during Operation Goldfinder, March 1977. Scott's arm, in a sling, was injured during the excavation (photograph 16).

Leonard Fiege, left, is interviewed by Dan Rather for the CBS show "60 Minutes.
Fiege, a former jet fighter pilot stationed near White Sands, passed an Army-administered lie-detector test of his claim that he found stacks of gold bars in a cave insid
Victorio Peak. During Operation Goldfinder, Fiege found that some time in the pa
the cave had been crushed in by Army bulldozers and sealed with iron doors. O
Noss, at age eighty-one, is assisted to the crest of Victorio Peak by Army personn
during Operation Goldfinder (photographs 17, 18).

Top, Sam Scott, a commercial airline pilot, and brother of treasure hunter Norman Scott, excavates a cave entrance at Victorio Peak. He is assisted by Colonel Charles Gunn (U. S. Army, Ret.), who was a member of the Bailey-Patterson group. Bottom, treasure hunter Joe Newman rests amid the ruins of the old Noss shack at the foot of the peak (photographs 19, 20).

Present at the Goldfinder Operation was Doc Noss's stepson. He survived a mysterious plane crash at Victorio Peak the day before Doc Noss was killed in 1949. In the
background, on the slopes of the peak, are some of the newsmen and treasure hunters
on hand for the twelve-day exploration (photograph 21).

timbering would have to be removed. This would have taken longer than the time permitted. . . ."[1]

On the fifty-ninth day, the miners broke through into a cavern at the end of the 218-foot tunnel. They had expected to find the other side of Fiege's cave-in. Instead, they had miscalculated because of a faulty compass. They had entered a small, unknown cave "undisturbed, covered by water-deposited silt."

On the morning of the sixtieth day, another cave was found. Miner Concepcion Villalobos attested later that "I personally entered a room or cavern in which dust lay upon the ground two to three feet deep which had sifted down prior to our excavation. . . . I was instructed to remove this dust by water slushing and after it had been removed to dig about the walls for a hidden passage to an alleged treasure room. I was prevented in completing the work by the Army, who at that time ordered us to vacate the area. Additional time was requested but the Army refused."[2]

On the final day, a Secret Service agent arrived at Hembrillo to witness the shutdown. He reported that "Loren Smith, exploration chief for the Gaddis Mining Company, [stated] that a horizontal tunnel had been excavated and they had gotten to the furthest point of previous exploration and were requesting a fifteen-day extension of the license . . . the Commanding Officer, White Sands, would not endorse this application. . . . The entire area was immediately reclassified as a secure area."[3]

The museum promptly applied for an extension through the Secretary of the Army.

Chester Johnson wrote a report that stated that only one purpose of the dig had been accomplished:

"It was the purpose of this project to prove or disprove the claim of Doc Noss that Victoria Peak is the site of the mine worked by Padre La Rue. It is the consensus of the project personnel, judging from the surface rocks, that this peak cannot be the location of the mine of La Rue. . . . There are no tailings or slag dumps on the peak and none is visible on the nearby hills.

[1] Chester Johnson, Jr., *Explorations at Victorio Peak* (Division of Research, Museum of New Mexico, 1963).

[2] Affidavit (July 26, 1976) in the possession of Joe Newman of El Paso, Texas.

[3] Communication from senior agent, Albuquerque, to chief, Secret Service (Oct. 1, 1963).

Any such should be obvious if they existed, as only 150 years have passed.

"The above leads to the conclusion that if a treasure was or is here, it was brought from some other location and stored here by persons as yet unknown. . . ."[4] The other goals, such as the locating of Fiege's cave or Noss's cache, were unresolved.

Before being released to the public, Johnson's report was censored by the Army. All mention of Army activity at the peak from 1958 through the end of 1961 was deleted. It was a section in which Johnson, quite accurately, described the Fiege discovery, the subsequent Top Secret excavation, the Army's denials of same, the production of the four affidavits, and the December 1961 event when "the four men returned to Victoria Peak and found that more work had been done since their previous visit on October 28. . . ."[5]

After nearly a year's delay, the museum's application to the Secretary of the Army was rejected on the grounds that it should go instead to the commanding general, White Sands.

Accordingly, the museum filed a new application stating "results of the exploration program . . . proved the existence of a number of open cavities within Victoria Peak similar to those described by the individual who claimed to have been in the caves and seen the artifacts and treasure. . . .

"One phase of the program consisted of a gravity survey, which indicated a large void or cavity, and a tunnel was started to reach this cavity. A total of 218 feet of tunnel was completed when the permit expired. In order to complete the project, an additional 250 feet of tunnel will have to be driven. . . ."[6]

The museum asked for ninety days to complete the work and assured that the "program can be conducted in a manner whereby it can be interrupted so as not to interfere with any range operations."

Attached to the application was a copy of a letter from museum director Dr. Delmar Kolb to Congressman Jake Pickle. It stated:

"Should some, or even a small part, of the items listed as pres-

[4] "Museum Report" (1963).

[5] From Johnson's uncensored rough draft, printed in the El Paso *Herald Post*, Aug. 29, 1973.

[6] Museum application to commanding general, White Sands (Mar. 11, 1965).

ent in the [Noss] room be recovered a wealth of historical materials would be available for study. Nowhere in the world is there a comparable collection from that period of Spanish and Mexican history. It is important, if the objects exist, that they come to a public institution rather than being dispersed as is reported the case for some of the items.

"True, the existence of such a treasure-trove is founded on evidence to which many people may take exception. However, the previous explorations indicate that at least some of the reports are founded on fact."

The above-cited application and letter were subsequently classified by the Army and removed from public files maintained by the Army at White Sands.[7]

The museum would persist for three years in its efforts to resume the dig. They were backed in the applications by Gaddis Mining, which already had spent one hundred thousand dollars but seemed willing to spend more in pursuit of the treasure.

The Army summarily refused entry. And with rather incredible cynicism, it misrepresented the museum's findings. In official press releases and briefings over the next thirteen years, the Army spokesmen consistently would state that no caves had been found, and that the museum and Gaddis Mining had concluded there was no substance to the treasure stories.

During the closing years of the 1960s, there arose two stories of gold bars at White Sands. The first came on November 3, 1965, when Ova Noss and her daughter-in-law visited Oscar Jordan at the land office.

Ova Noss said that she had heard rumors "from so-called reliable sources, that a high officer in the government, they believed it was a retired general, was involved in activity around the alleged treasure cave, or in the vicinity. They stated that they had also heard rumors that gold bars were being taken out of the cave and being disposed of through [a banker] in Lubbock. . . . They also stated that someone had taken some coins out of the cave. . . ."

In view of Ova Noss's track record on well-founded complaints, the land office investigated but was unable to substantiate the rumors.

A few years later, museum archaeologist Chester Johnson en-

[7] The author's copies were obtained from the Museum of New Mexico.

countered a second White Sands story. He found it while on a project in Korea. A serviceman he met in Korea told him "that he and some other servicemen had found a cache of gold bullion on the missile range east of the Organ Mountains. . . . they had buried the bullion under a pickup truck and covered it over—truck and all—and did not know what to do about retrieving it."[8]

The site, east of the Organs, is the San Augustin Pass area, which would be included in the 1976 exploration applications.

The 1960s ended as they had begun, with rumors and unresolved claims. The 1970s would see an acceleration of the treasure drama. For the first time famous personalities would enter the play. Some would have lead roles, others having only bit parts. They would include lawyer F. Lee Bailey, White House counsel John Dean, and White House domestic affairs czar John Ehrlichman.

At the center of the drama, however, would be an obscure player named Fred Drolte of Dallas, Texas.

What is known about him is that he has one arm; is a pilot; and in 1966 he was convicted of conspiracy to smuggle 371 Army carbines and 100,000 rounds of ammunition to revolutionaries in Mexico.

He is a man surrounded by mystery. His lawyer once described him as a "stockbroker." The Dallas city directory listed him as being in the oil business. When he came out of prison, he had no visible occupation. He was rumored to have strong CIA connections. He settled in El Paso and seemed to be well financed. He hired bodyguards to protect his home and family, from what he would never say. And he used armed guards to seal off a seldom-used public road in the Caballo Mountains where, in 1976, he commenced secret mining operations.

[8] Howard Bryan, "Army Censors Report of Military Probe" (El Paso *Herald Post,* Aug. 29, 1973).

XIV

The One-armed Man Is King

Fred Drolte was adverse to my seeing the gold at the first meeting. Without him, of course, I had no guide to the specific cave entrance. And besides, there were men at the peak with high-powered rifles. Drolte's team. That's why I didn't go out alone. I had no white flag to get past those rifles.

—F. Lee Bailey interview

F. Lee Bailey, attorney, born in Massachusetts in 1933, is five feet, seven inches—shorter than his reputation leads one to expect. He is stocky and fights his weight. But his movements are quick, his voice strongly authoritative, and the overall impact is of a very tough, very intelligent *competitor*. He likes to compete. It is a passion that, most likely, is the basis of his success and power and riches.

He is a romantic character, suitable for a novel or a television series. He is sophisticated, witty, well tailored, good-looking, and flies his own jet from adventure to adventure. During the 1960s and 1970s he was defense counsel for such clients as Dr. Sam Sheppard, the Ohio osteopath who spent ten years in prison for a crime he didn't commit. Bailey was also the defense for Albert De Salvo, the Boston Strangler; for the defendants in the Great Plymouth Mail Robbery; for Captain Ernest Medina, the soldier accused of responsibility for the My Lai massacre in Vietnam; and for Patty Hearst, the kidnaped heiress.

Bailey, always on the move, nevertheless stopped long enough to have a conversation with a one-armed man, to see a bar of gold, and to take it to Washington. There, Bailey presented a practical, feasible plan to legally extract the gold from White Sands. It was a plan that made sense and was fair.

It was never accepted. Instead, F. Lee Bailey, master lawyer, nearly got the blind staggers after encountering a bureaucratic maze where everything was curiouser and curiouser. Dynamic inaction, decision postponement, and creative status quo were the policies of the day. To Bailey, to the White House, to the Congress, and to the press, the bureaucrats presented a stone wall. The bureaucrats were indifferent to all but the welfare of themselves and their agencies.

All that would take place in Washington.

In New Mexico, men patrolled desert roads with guns, and the one-armed man was king.

The renewed interest in White Sands gold actually commenced in November 1968 with the publication of "The Incredible Victoria Peak Treasure" by Xanthus Carson in *True Treasure* magazine.

The article began like a hundred-yard sprint:

This is the story of the granddaddy of all New Mexico treasures —a $350 million cache that Chiropodist Doc Noss reportedly stumbled upon while deer hunting back in 1937. Eyewitness accounts support his incredible find, and he was murdered over 51 gold bars that presumably came from the trove. But if this immense underground treasure really existed, then the bulk of it is still there, sealed away within a mountain inside a closely guarded military reservation—perhaps forever!

Mr. Carson then proceeded to tell the story, from 1937 through the museum dig of 1963.

It aroused some attention. Several parties of treasure hunters got their blood up and prowled the area for a few months. One such party sought out Harvey Snow to be their guide.

Snow is in his fifties, with thick, curly gray hair; tall, tanned, lean-muscled, he has a square jawline and a face with lines like sculpted oak.

For twenty-five years he has been exploring all of the White Sands region—from the southern Organs to Trinity Site. He has done it afoot. With little formal education, he has an exceptional vocabulary and is widely read in both the sciences and the humanities.

Snow believes that the treasure of Noss is in Hard Scrabble Mountain, twenty miles north of Hembrillo.

In the 1950s, a cowboy working the San Andres had told Snow this story:

Many years earlier, in the 1930s, the cowboy had gone up early one morning to search Hard Scrabble for stray cattle. As he approached the mountain, just after daybreak, he saw a man leading some mules up the mountain. The mules were carrying packs, but the packs were empty. The man was Doc Noss.

The cowboy had heard of Noss and his dangerous reputation. Prudently, he hid himself. To his amazement, he saw Noss disappear halfway up the mountain into a nearly invisible cave entrance.

Intrigued, the cowboy spent the day in the area, never wandering far from where he could see the cave. At dusk, just before nightfall, Noss emerged. The mules now had loaded packs, and Noss led them south, toward Hembrillo.

The cowboy waited about one hour, until he was confident that Noss had cleared the area. Then he went to the cave.

It was night, and the floor of the cave sloped sharply downward. The cowboy lit matches and found it had been cut with crude steps—identical to rock steps cut by Indians on the trails leading up to mesas.

"He told me he did not go down all the way," said Harvey Snow. "He was scared and he never went back there. He was afraid Noss would find out and kill him."[1]

Snow had made several attempts to find the cave but was unsuccessful. The years passed, and Snow concentrated on working his own silver lode in the Organs. Then, in late November 1968, following publication of the Carson article, Snow was approached by three area ranchers. They asked Snow if he could lead them past the Army patrols to the Noss site. Snow replied by telling them the story he'd been told by the cowboy.

[1] Interview with author.

"There were no patrols at Hard Scrabble, though it's on the reservation. It would mainly just be a matter of moving at night."

Using the cover of arroyos, they drove in from the Jornada as far as they dared, then hid their Jeep with rocks and brush. They hiked the rest of the way at night to Hard Scrabble.

"We spent two full days there looking for the cave and couldn't find it. We had no more supplies, so we went home."

Snow's interest was peaked, however, and he suggested they make a second try. This time, only two men would travel. Snow would be dropped at Hard Scrabble, make camp, and search the southern face of the mountain. It was arranged that he would have three nights and three days to search and would be picked up on the fourth night.

Harvey Snow's description of what followed is hard to believe. It is too romantic, too incredible. In addition to ricks of treasure, he will describe an east–west cavern system across the Jornada, a claim that the New Mexico Bureau of Mines labeled "geologically ridiculous." (The bureau's senior geologist, Robert Weber, conceded, however, that a *north–south* system down the San Andres, also described by Snow, was possible and even probable.)

On the other hand, Snow has produced artifacts from the region, including swords, rifles, and other equipment dating from Spanish times to the late nineteenth century. Furthermore, he has a reputation for trustworthiness and veracity. For better or for worse, here is his story:

"On the second day, I found the cave with the sloping steps. I went down the steps; down and down. I don't know how far. I estimated maybe thirteen hundred or fourteen hundred steps. The bottom step, the last one, was rounded at the bottom so that when you stepped on it, it would roll. It was tied to a bow and arrow with rawhide, but the rawhide had rotted away long before I got there.

"At the bottom you are in a big room. There is a stream of water running through. Now Noss described a cold stream, but this stream is hot. It has a copper and sulphur taste. It ran from east to west, toward the Jornada, and along what I assumed to be an old earthquake fault.

"I followed it, going from room to room. In many places I had to get down on my hands and knees, and in a few places on my belly. After that first room, where the steps come down, I came

into another room. Here I found some things. I found small stacks
—one of gold, one of copper, and one of silver.

"I figured I would come back for that and went on. I next came
to a big room. Here there were a bunch of side tunnels running
north and south. They were all natural, nothing man-made. Here,
where they intersected, they made a big X. I did not go down
those north–south tunnels. I stayed with the stream, going west.

"At the far end of the main room I found some things I cannot
tell you about. But I will tell you what happened next. I kept
going west, kneeling, crawling, and walking, hauling my little pack
of food. I kept going through there for two days and two nights. I
eventually followed that tunnel fourteen miles.

"Where I was coming out was under the Jornada. That tunnel
must not be far below the surface, because I could hear jets flying
overhead, and when the train was traveling it sounded so near you
thought it was in the tunnel with you.[2]

"Finally, I felt some fresh air on my face and then I saw some
light. The tunnel had been getting narrower and narrower all this
time, and I figured I was about to the end of it. I came to the hole
where the light was coming from and stuck my head through. I
was standing in a hole covered by bushes and I was smack in the
middle of the Jornada."

Snow would never say what he found in the main room. He
would not say if it was treasure or bodies or what. He said he subsequently made a "few trips" to the room where he found the
small stacks. He *indicated,* or at least did not deny, that he may
have removed some of the silver bars.[3]

Snow's discovery, whether true or false, marks the beginning of
the modern chapters on White Sands. It is Snow who will be
sought out by Drolte in the 1970s as the man who best knows
White Sands.

For the next episode, we go to Washington, to examine an exchange of memos between the White House and the Treasury Department in the summer of 1970.

John Ehrlichman was Richard Nixon's chief of domestic operations, No. 3 in the White House hierarchy after Nixon and Bob

---

[2] The Atchison, Topeka, & Santa Fe has a track running north–south on
the western edge of the Jornada.

[3] Interview with author.

Haldeman. On July 9, 1970, Ehrlichman wrote the following memo to David Kennedy, Secretary of the Treasury:

"While I was with the President in San Clemente last week, I was called on by a Mr. Keith Alexander, who was introduced to me by a County Assessor from Northern California named Earl Corin, whom I have known for years.

"Mr. Alexander claims to have secret knowledge of the location of 742 bars of gold weighing between 40 and 80 pounds each. He obtained knowledge of the whereabouts of this cache from an aged Indian who assertedly is the last of his tribe. Some of the gold was mined by the tribe and some dates back to Aztec and other ancient times.

"Mr. Alexander seeks agreement of the federal government that he be permitted to go to the location of the gold, possess it, and move it to a United States Mint for sale to the government at the controlled price without jeopardy.

"He contends that without prior agreement in advance from the government mere possession of the gold would be a violation of statute on his part.

"For now he requests that we put him in touch with a representative of the Treasury, or other competent federal official, in order that he can negotiate a proper agreement with the assistance of his attorney.

"I would appreciate your advice as to the proper way to proceed with Mr. Alexander, who appears to be a reputable geologist but for whom I cannot vouch. He is a former resident of Phoenix, Arizona, now residing in Northern California. His original home was in Salida, Colorado."

It would later be learned that Keith Alexander, while working in the White Sands area, had been told the story by Benny Samaniego.

Unexpectedly, Secretary Kennedy responded to Ehrlichman that the finding of gold presented international problems:

"Since the agreement of March 18, 1968, established the two-tier market, it has been firm U.S. policy to purchase no gold whatever except from foreign monetary authorities. We have worked hard to secure strict adherence to this policy from the Group of Ten and all IMF member governments, and of course have been meticulous in setting a good example ourselves. The Treasury could not, therefore, under any circumstances purchase the alleged cache of gold. . . ."[4]

However, Kennedy offered a reasonable alternative:

"No license or other authority from the federal government is required for any individual to search for, discover, and possess buried treasure." The finder was required only to promptly report his discovery to the Office of Domestic Gold and Silver Operations or to the Secret Service. "The report will be investigated and if the treasure is authentic the finder will be authorized to dispose of it in the private market."

Negotiations on the Alexander application continued into the fall, but after the initial communication between Ehrlichman and Kennedy the matter passed into the hands of the professional civil servants at Treasury. Once the bureaucrats had control, they reversed the Secretary's firm assurance that the *finder will be authorized to dispose of it in the private market.*

On September 11, 1970, the Treasury Department drafted a reply to be sent to Alexander over Ehrlichman's signature. In effect, it said: *Trust us, show us the gold, and we will let you know how much you can keep, if any, and how much you can sell, if any.*

The pertinent part of the memo stated:

"The Treasury Department has advised, however, that no advance commitment can be given regarding . . . the matter of the disposition of the gold. Decisions on these questions can only be made after a full disclosure of all the facts. If after investigation it is determined that gold has been discovered as part of a buried treasure concerning which the holder has not been guilty of any aggravated violations of the Gold Regulations, and the owner of the gold cannot be located, the Treasury Department will consider authorizing the sale of the gold to a licensed refiner."

Upon receipt of that information, Alexander temporarily abandoned his official negotiations with Washington. He returned to New Mexico, where he met Pat Patterson, a helicopter pilot familiar with the Fiege find, and Fred Drolte.

"It began in February 1973," said F. Lee Bailey. "I was in Kansas City with a case. A former client, George Milano of

[4] Treasury memo from Secretary Kennedy to John Ehrlichman (July 17, 1970).

Pueblo, Colorado, came to see me there and asked if I'd meet with some people concerning a gold treasure. I had two meetings and was introduced to Pat Patterson of Albuquerque and a Fred Drolte of El Paso. They said they represented a group of approximately 50 businessmen who needed legal advice and assistance in recovering a vast treasure from White Sands. I gathered that two men—Keith Alexander and Fred Drolte—had actually seen *staggering* amounts of gold. They described it as being cast in ingots averaging 70 to 85 pounds each. They said there was also a vast amount of gold coins and other valuable artifacts. Some of the treasure was buried in caves and other underground.

"Alexander, Drolte, and Patterson each had arrived at their knowledge of the treasure by independent avenues.

"They believed the bullion to be of ancient origin, and an estimate of the total amount ranged between 50 tons and 110 tons.

"They convinced me to the extent that the following day I phoned John Mitchell, the former Attorney General. He made an arrangement for me to meet with [John] Ehrlichman at the White House."

At a second meeting in Kansas City, Bailey was shown a terrain map of the Hembrillo region.[5] The treasure sites were Section 16, which is Victorio Peak; Section 22, immediately southeast of Victorio Peak; and Section 9, which is Geronimo Peak.

The main treasure was under Geronimo Peak, not Victorio.

The information of the sites came from three primary sources: Drolte's claim to have actually seen the room; Alexander's descriptions; and Patterson's detective work over the years, which had uncovered correspondence between two of Noss's workers, Benny Samaniego and Dick Richardson, both deceased.

Drolte told Bailey that he had descended through the hole at the top of Victorio Peak, crawled through a fissure created by a fault, and thence into a long cavern, which allegedly extended for some distance north to a point somewhere under Geronimo. Drolte gave much the same story as Noss as to what he found there: stacks of gold bars; old Wells Fargo chests; and skeletons.

Bailey told the story during a four-hour, afternoon-to-evening interview at the Royal Orleans Hotel in New Orleans' French

[5] Apparently the map he saw was "Kaylor Mountain, New Mexico" map sheet 4649-I, Army Map Service, Corps of Engineers.

Quarter. He was interrupted frequently by aides and by phone calls concerning cases of his throughout the country. His demeanor had its contradictions. Overall he was calm, spoke with authority, and—as later government documents would show—with precise truthfulness. In contrast to the calm, however, he drank at least half a dozen bloody marys and smoked cigarettes with a frenzy.

"Was Captain Fiege involved in any of the meetings?" he was asked.

"Fiege? No. He would align himself with the second wife of Doc Noss [Mrs. Violet Noss Yancey of Fort Worth] and the Scott brothers of Florida, who acquired their power of attorney. But they don't know where the gold is. They know the general area, but Fiege can't find the gold. And Ova Noss can't either. My clients have seen it and touched it in the past eighteen months."

"What happened after the Kansas City meetings?"

"I phoned John Mitchell. . . ."

"You didn't have lunch with him?"

"No, I've never even met John Mitchell. *After* I called him, John Mitchell had lunch with John Dean and told him about it. That's how the story of the lunch got into Watergate."

"How did Mitchell respond to you?"

"Mitchell referred me to John Ehrlichman. That was in early March 1973. I met with Ehrlichman at his White House office and told him, 'This thing can be a big thing or a lot of bullshit. My clients say the gold is there and want a legal means to remove it from a federal reservation. But before we waste time, let me see if it's really there. If you give me the authorization to proceed without Army knowledge, I'll look at the gold, touch it, photograph it, and return with a sample.'"

Without Army knowledge. Bailey from the beginning was suspicious of the Army's good intentions.

"What did Ehrlichman say?"

"He gave me verbal authorization to proceed. He said to work out details with his assistant, Tod Hullin, and Ron Brooks of the Treasury Department."

A White House memorandum dated March 29, 1973, and signed by Tod Hullin states:

Per Mr. Ehrlichman's instruction, I met with F. Lee Bailey this afternoon. He indicated he was representing a group of approximately 15 men who knew the whereabouts of approximately 90 tons of gold and 10 tons of silver (conservative estimate). Some of the gold was in the form of pieces of art. The gold cache is split into two parts: the first part is approximately 11 tons and is immediately available, and the second part is the remaining 89 tons which could be made available in 72 hours. The gold is located on government property. It is believed that there may be a "mother lode" associated with this find that is of undetermined value. . . . It is said to be located somewhere near the Mexican border. . . .

I indicated that the major question was whether or not a cache of this sort or any cache of gold existed. Mr. Bailey agreed and asked if the following plan of action would be appropriate:

1. Mr. Bailey write a letter to his client indicating that he (Bailey) would fly to the cache and take some pictures of it and two or three random samples of the cache to be given to the government for testing.

2. Mr. Bailey would return with these samples to Chicago (where he is trying a big suit) and would turn these samples over to a designee of the Department of the Treasury to be returned to Washington for testing.

If the tests prove positive, Mr. Bailey would ask that we move rapidly to put this entire cache under the government's control for appropriate disposition. He is convinced that this group has no criminal intent and is trying to do the right thing.

Mr. Bailey is concerned that this find remain highly confidential until it is in the government's control.

I indicated that once the government had tested his random samples and had determined that there is a reasonably high probability that a large cache exists, that we would be ready to ask the Treasury Department to move rapidly.

Mr. Bailey departed and indicated that he would probably fly to this location this coming weekend (March 30, 31, April 1) and that he would like to bring this to a conclusion as rapidly as possible.

Additionally, Mr. Bailey indicated that he was surprised to learn from Mr. Ehrlichman that the government had a firm agreement with some foreign countries that we would not acquire any gold from the private market. He indicated that Ehrlichman had hinted that this may change sometime later in the year.

That weekend Bailey flew his Lear down to El Paso and met with the principals.

"Was Drolte present?"

Bailey stirred his drink. "No, Fred didn't want to come to the meeting room. He was nearby, calling the shots. Pat Patterson was the main principal in the room."

"Why didn't Drolte come?"

"He had expected me to come from Washington with a written *carte blanche* to recover the gold—immunity from past trespasses, a license to dig it up and keep it, everything. When he found out I had only oral authority to go in and pick up a sample, well, he became adverse."

"So you didn't go in to Hembrillo?"

"No. Drolte was to be the guide. And there were people, his people, patrolling the peak with high-powered rifles to discourage visitors. That's why I didn't go out alone. I had no white flag to get past those rifles. Within a week, Patterson came to Chicago and gave me a sample bar taken from the site. I sent it to the Treasury Department, and they assayed it at 60 per cent or 65 per cent, something like that."

A memorandum from Treasury official Ron Brooks to the Secretary of the Treasury dated May 15, 1973, states:

Mr. F. Lee Bailey has been in contact with a group of persons who allege to know the location of a very large horde of gold and to be located on a U. S. Government military reservation in New Mexico. The gold is alleged to have come from Mexico and is supposedly church gold or Indian gold. . . .

Mr. Bailey went to New Mexico to interview the group and to view the gold, but was denied access to it because the group became suspicious that Mr. Bailey may have been followed by federal agents.

Mr. Bailey supplied [this office] with a small bar of gold for analysis. This bar, which was analyzed by the Bureau of the Mint,

was shown to consist of 60 per cent of gold and 40 per cent of
copper, and had apparently been prepared by an antique
method. . . .

Bailey met with Brooks the following day and was told that the
bar, weighing slightly more than one pound, had proved to be
gold and was apparently ancient. Bailey smiled. Brooks then
jolted him with Treasury's decision. Despite the assay, the department would not agree to Bailey's conditions of secret entry, nor to
immunity from previous trespass.

The cause of Treasury's decision is not specifically spelled out
in the memos. It was, however, partly based on the fact that the
secret agreement with foreign governments prevented the United
States from acquiring domestic gold; partly on a suspicion that
Bailey was somehow leading the government into a complicated
gold plot, the nature or intentions of which were unknown; and
finally, sheer bureaucratic inertia had a lot to do with the decision.

Bailey was stunned. He felt betrayed and began to suspect that
someone in government was already getting to the treasure.

In the ensuing month, Bailey attempted on three occasions to
compromise on the conditions, but without success. Finally, on
May 15, the gold bar was returned to Wayne Smith, Bailey's
Washington representative.[6] Smith was handed a letter from
Brooks stating:

"Dear Mr. Bailey:

"I am returning the material you provided for me. . . . As I indicated to you on the telephone, the Treasury has no interest in
proceeding further with this matter."

Bailey tried another avenue and presented an amended plan to
Thomas Wolfe, director of the Office of Domestic Gold and Silver
Operations.

Bailey presented the background and his legal interpretations of
the law on treasure. He added:

"My clients specifically propose . . . that I, as their lawful
agent, be acquainted with the physical location of a portion of the
more than 100 tons of the gold they wish to deliver; and that I

---

[6] The bar was returned to Smith by a Secret Service agent. The Service a
few days later reversed itself and seized the bar as contraband. As of 1977,
the service had neither returned the bar to Bailey nor compensated its
owners.

can then call a duly authorized agent of the Treasury Department and ask him to come pick it up, accepting a receipt for the same. If after suitable examination, it is determined that this is not stolen property, it would be returned to me as agent and trustee and could be sold at market prices. . . .

"To establish their credibility, these clients propose an initial delivery of 292 bars of about 80 pounds each, approximately 750 fine. . . .

"I have [recently] reported to the Treasury that part of the gold has been stolen, and there is reason to believe that more can be. I am also informed . . . that there is currently a group from Texas which expects to take, with government permission, a portion of this gold on Sunday, June 3, 1973, and remove it. . . ."

Bailey asked for a swift response.

On the same day, at White Sands, an Army spokesman told columnist Jack Anderson that important missile research could not be interrupted "every time someone comes to us with a gold gleam in their eye . . . so far, no one has produced gold."

Bailey's gold bar was ignored. It didn't exist. "It was like Alice in Wonderland," he said.

Jack Anderson published a story on Bailey's dilemma the following day. In view of the gold crisis then prevalent in the Western world, the government's refusal to at least pursue the claim became doubly curious.

On June 25, 1973, the matter was introduced into the Watergate hearings by John Dean, who briefly described the contact between Bailey and Mitchell.

The Watergate testimony triggered national press interest. It also alarmed the New Mexico state land office, which immediately informed federal authorities of its long-standing legal position regarding gold at White Sands.

The Army in the meantime intensified its security at Hembrillo Basin.

Bailey continued to try new doors. On November 15, 1973, he signed a contract with the state of New Mexico whereby the state would share in 25 per cent of the treasure in return for its assistance in gaining entry to the site.

Twelve days later, someone in Bailey's group beat the security and penetrated Hembrillo Basin.

"On November 26, 1973," said a Secret Service report, "word was received that persons unknown had entered the White Sands Proving Grounds sometime over the Thanksgiving weekend. [name censored] stated that today he had received an intelligence report which indicated that someone had entered the range and that evidence of dynamiting had been discovered in the area of Victoria Peak.

"Hobbs said that the persons responsible apparently had a key to a gate on the west boundary of the range; had knowledge of the helicopter surveillance being curtailed because of the fuel shortage; had knowledge that the rangers had the holiday off. He said that the FBI had been advised and that an investigation is underway to determine the identity of the person on the inside responsible."[7]

The Secret Service report strangely minimized the intrusion. A slightly fuller report was prepared by Army security at White Sands. It states:

"On November 26, 1973, information received from Mr. George W. Brazier, Jr., deputy for Installation & Housing, Office of the Assistant Secretary of the Army, indicated that sometime during the Thanksgiving weekend, an intrusion was made into the Victoria Peak region. Briefly it was reported that a group had entered the range, used dynamite, and removed 37 tons of gold and artifacts.

"Immediately two range inspectors were dispatched to the Hembrillo Basin to investigate this report. . . .

"Their findings substantiated the report to the extent that several blasts were set off at the base of a shale outcropping approximately one mile east of Victoria Peak at a place known as 'the Cottonwoods.' Additionally, there was evidence of a campsite and 'sign' indicated an estimated five to nine personnel with two or three vehicles, one with a small trailer, had been in the area. To cover their tracks upon departure, brush had been dragged behind the trucks, and the trespassers had in some manner opened the west gate, as it was evident that the intruders went out the gate.

[7] Secret Service memo (Nov. 27, 1973).

No evidence was found indicating that any gold or artifacts were removed. Inquiry into this matter is being continued by the FBI.

"As a result of this incident, a trailer has been located near the west boundary of the range at a road junction where all vehicular traffic can be intercepted. This site is occupied seven days a week, twenty-four hours a day by range inspectors and military policemen. Also, a gate has been installed on the road leading into the basin from the east. Since applying these security measures, no incidents of a significant nature have occurred. . . ."[8]

The combined reports of the Secret Service and White Sands raise more important questions than they answer. Why did the original tip come from an Army deputy in Washington, rather than from White Sands' own security? Who ordered the grounding of the helicopters and the suspension of patrols, civilian and military? How did the intruders obtain a key to the west gate? Why in the face of an apparent conspiracy, and possible bribery of government personnel, hasn't the Secret Service, the FBI, or the Army pursued the investigation?

The investigation in fact went no farther. Furthermore, the strengthening of security measures had no effect on a subsequent illegal entry in the summer of 1974. That, too, went uninvestigated.

Negotiations for legal entry by Bailey continued through the spring of 1974. In May, Army Secretary Howard "Bo" Callaway instructed his staff to draw up plans allowing a controlled entry and search of the site. The project was code-named "Goldfinder."

The Army's co-ordinator was Bland West, deputy general counsel. West would respond with positive interest and would henceforward function as a wise and fair umpire among the conflicting and often mysterious interests of the various government agencies and the claimants.

West's first task was to sift the various claims and come up with a list of recognized claimants.

West's list:

•*Bailey's clients*—sponsored by (the state of) New Mexico, which has entered into a contract with Bailey under which it consents to the search of Section 16 for treasure-trove on land owned

8 "Security Office Briefing on Hembrillo Basin-Gold, 5 March 1974." The document is in the possession of the judge advocate general's office, White Sands.

by the state, in return for 25 per cent of any trove found. The state plans to exact similar agreement from other claimants seeking trove on Section 16.

•*Shriver group*—Shriver was one of the "detectives" who had tracked down a potential site of the treasure and then had obtained a General Services Administration contract to search for treasure on the White Sands Missile Range. The contract, said West, was ineffective because the Army never agreed to it.

•*Roscoe Parr*—Noss's old-time partner, who buried Doc. He told West that Noss wanted him to "succeed to his right to the treasure. Parr is sponsored by House Appropriations Committee chairman Mahon.

•*Ova Noss*—wife of Doc Noss.

•*Violet Noss Yancey*—Doc's second wife, who had a contract with professional treasure hunter Norman Scott.

In May, West invited the claimants to meet with him in Washington. The meeting was held May 23, and those present included Drolte and his wife; Patterson and his wife; David Norvell, attorney general of New Mexico; West; and Bailey.

In an Army memo prepared on that date, West wrote:

"Drolte and Patterson are the principals in the Drolte enterprise. Bailey has a 10 per cent interest, with Drolte and Patterson dividing the balance on what I believe to be a 60-40 basis. However, these two have disposed of fractional interests in their shares to other persons who are named in a letter from Drolte to [me]. Drolte also recognizes overriding interests of the state of New Mexico and the United States Government to the extent of 25 per cent each of the value of any gold treasure found. . . .

"Prior to this meeting, the Drolte group apparently had endeavored to absorb or eliminate other potential treasure seekers so that no one else would be left in the field. To this end, they excluded the Noss widows from consideration for lack of capital to participate in the operation; they planned to stigmatize the Shriver group as disreputable and unqualified for an entry permit; and hoped to absorb the Roscoe Parr interest by persuading Chairman Mahon to exert his influence for that purpose. This approach was not used at the meeting. Instead, Drolte, Bailey . . . and others argued that they should be allowed to go into WSMR first for two reasons. . . ."

The reasons were that Drolte was the only man "who knows where the treasure is and can furnish convincing, detailed facts concerning its existence, including a gold bar, and because they alone have a contract with the state of New Mexico. . . ."

West described Drolte as "a tall, well-built, gray-haired, vigorous man who appears to be in his sixties. He revealed little about himself except to say that he had been a miner and in the oil business. He spoke more than anyone else during the meeting and several times talked in a rather rambling fashion about the lost gold and White Sands, the tortuous way down into the ground under Victoria Peak to the caches of gold located there, the dangers involved, including a store of ancient and unstable dynamite located along the path to the cavern, the danger of cave-ins, and the possible origin of the gold as coming from the Aztecs in what is now Mexico some 300 years ago. He also spoke, without seeming to notice any inconsistency, about the possibility that the source of the gold was a vein deep within the peak and that the gold had been mined, smelted, and made into gold bars all within the mountain.

"Mr. Drolte speaks with apparent candor and sincerity, but much of what he says about the gold is so contradictory and incredible as to suggest that he is not rational on the subject."

As for Bailey, West said he was "a criminal lawyer of national reputation [who] seemed out of place in this group of treasure-seeking visionaries. At times he appeared uncomfortable during Mr. Drolte's ramblings about the lost treasure or when an intemperate or inaccurate statement was made by some other member of the group. He spoke of the possibility of there being a huge vein of gold deep in Victoria Peak, complementing the statement to that effect by Mr. Drolte."

West added: "While the meeting confirmed the fact that Bailey has a 10 per cent interest in the Drolte enterprise, the long-abiding suspicion that Bailey is playing some other game remains unanswered."[9]

West concluded the meeting by spelling out a tentative plan for a forty-eight-hour entry. The rather simple and fair original conditions of Goldfinder, however, would later be made complicated by

[9] "Memorandum for the Record" dated and signed by Bland West (May 23, 1974).

personnel at White Sands and eventually would be turned down
by the claimants. This is a consistent pattern in the federal performance. The upper echelons, closest to the White House and the
Cabinet appointees, approve sensible plans to resolve the questions, only to have them reversed by bureaucratic functionaries.

In the New Orleans interview a year later, Bailey was asked
about Drolte. Like West, Bailey could never learn very much
about the man.

"I don't know much about him," said Bailey. "He is an engineer, in his late fifties. . . ."

"What about his indictment for arms smuggling?"

"That seems to have been treated very lightly by the government," said Bailey. "They accepted a plea of *nolo contendere* and
little jail time, as I understand."

"How did he get knowledge of the gold?"

"One of Doc Noss's old employees told him about it. But the
old Noss entrance had been dynamited shut. In the 1950s, Fred
found a different entrance—sort of a quasichimney going almost
straight down. He went in and found the gold."

"Who else has seen the gold besides Drolte?"

"Well," and Bailey hesitated, "an El Paso man named Joe Newman. He was unknown to us, unknown to Drolte, but in 1973
Drolte picked up information that this Joe Newman had been
arrested by the Army for trespassing at Victorio Peak. So Fred
contacted Newman. . . ."

(Army records of 1975 stated that a Joe Newman had been
arrested at the peak in October 1973. The report, however, was
false, and the document later disappeared from White Sands files.
Newman in fact was never arrested nor caught on the White Sands
reservation.)

"Had Newman been into the cave?" Bailey was asked.

"Yes. Somehow Drolte persuaded Newman to describe the
cave, and the description fit. Newman said—and this impressed
me—Newman said that after what he saw he couldn't sleep for
fourteen days."

"What did he see?"

"He said he saw a huge pile of gold, and 'the pile went as far
back as I could see with my flashlight.' Just then he heard noise at

the mouth of the cave. He climbed out and was headed away from the area when the range riders arrested him.[10] A few months after that, our group went out to see if the gold was still in place."

"Was it?"

"It was."

[10] The range riders, of course, didn't arrest Newman. They apprehended his partners. See Chapter I.

XV

Joe Newman

Governor King of New Mexico was presented a "gold bar"
recently by the Drolte-Bailey group, which has been active in
Santa Fe for the past three weeks, trying to get Governor King to
pressure the United States Army into granting the said group the
first opportunity to locate the alleged lost treasure. . . . It
weighed one pound. . . . The [Los Alamos] laboratory assay
learned that the content was 63 per cent gold, 33 per cent copper,
1 per cent silver, and that the remainder was composed of tiny
fractions of other metals. They were unable to determine the
source of the bar or its age. . . .

> *—"Memorandum for the Record" dated and*
> *signed by Bland West, deputy general counsel,*
> *Department of the Army (Aug. 7, 1974)*

The bar referred to was the second gold bar produced by Bailey
and Drolte in their effort to gain Army approval for entry. This
time, having in mind that their first bar had been confiscated by
the Secret Service, they dealt exclusively with the state of New
Mexico. The bar was delivered on July 25, 1974, and after the
assay proved it to be gold, it was exhibited at a press conference
in Santa Fe.

Bailey obtained the bar from Fred Drolte. Drolte, in turn, may
have obtained the bar from Joe Newman. The story was told by
Newman on August 5, 1976.

The temperature in El Paso was 107 degrees. The sky was high,
clear, and cloudless, and the sun was so unshielded that a person
could burn lobster red in six hours. In the shade, however, it was

not at all unpleasant. El Paso is at nearly 4,000 feet elevation, and the air was breezy and dry.

In downtown El Paso, there is a carpet and rug shop owned by Joe Newman. His Citizen Band-radioed Jeep is usually parked outside. But on this particular Thursday it was gone. Joe Newman was on Interstate 25 headed north to the Upham cutoff where he'd turn east, cut across the Jornada, and sneak into the San Andres.

As he drove, he pointed out places important to him as a boy and as a man. To the left stretched a band of green land, an irrigation belt sharply distinct from the brown and red land of southern New Mexico. Extending as far north and south as the eye could see, it was the Rio Grande Valley, bisected by the famous river itself, a river large in legend and length but surprisingly tiny in size compared to such wonders as the Missouri, the Ohio, or the Mississippi. It was shallow and muddy, "too thick to drink and too thin to plow."

As he drove, Joe Newman passed Mesilla, once the seat of Dona Ana County, the pride of southern New Mexico until it was bypassed by the railroad in the 1880s. Mesilla had been the home of Joe Newman's grandfather.

Immediately to Joe's right were the Organ Mountains, a fairytale range barely twenty miles in length but enchanting in aspect, with giant columns and castlelike turrets. Joe Newman had crossed them and crisscrossed them half a dozen times on foot. "But there are so many hidden valleys," he said, "so many caves and tunnels and ravines that you could hide an army in there and not find it in a dozen years. In fact, when my father was a boy, a whole tribe of Indians used to come down from there at Christmastime to see Mesilla. They lived up there somewhere in the Organs, but nobody could ever find them."

Beyond the Organs is the Upham exit. There Joe Newman left the interstate and onto a bulldozed road leading into and across the Jornada. He drove at fifty miles an hour, for the Jornada was dry and flat as a tabletop, uninterrupted save for the prickly pear, the mesquite, and the cactus.

As he drove, Joe Newman pointed to the left, to the Caballo Mountains. In there, he said, Fred Drolte was digging a dummy mine. "He's staked out a claim up there. There's nothing on it," said Newman, "but it gives him an excuse to bring mining equipment into the Jornada. He wants people to think he's mining the Caballos when in fact he's sneaking into Hembrillo.

"I was up there two months ago. We're supposed to be partners, you know? I had heard he was in the Caballos, and I drove up to the site in this Jeep. The road was blocked by barricades, and I was met by two armed guards. I had my .38 in my lap, and we were all just kind of staring at each other when Drolte drove up. He didn't say 'Hello,' he didn't say 'What you doing here?' The first thing he said was, 'Joe, I haven't been up at Vicky Peak.' He was real nervous and he repeated it, 'Don't ask me anything about Vicky Peak.' He was so nervous I figured I should get out of there before trouble started. But before I left I told him that whoever went in to Vicky Peak was in trouble because their entry could be proven by the other claimants.

"I think maybe Drolte has worked a deal with the Army to sneak in there and carry something out."

Joe Newman had first heard of treasure in the San Andres from stories told to him by his father. He, his father, and his older brother Robert had spent much of the summer of 1939 prowling the Organs and San Andres looking for the La Rue mine. Joe was ten years old, and it was one of the best summers of his life.

"I never forgot the legend, and in 1972 a group of us got together in El Paso and decided we would do some more looking.[1] We didn't expect to find anything, I guess. But it was a hell of a good excuse to get out in the mountains by ourselves."

The group began walking at the southern edge of the Franklins overlooking El Paso. For nearly every weekend of the ensuing six months, they walked northward, from the Franklins to the Organs, eventually into the San Andres. "We were looking for old mine shafts," said Joe, "we were looking for the Lost Padre mine, the La Rue mine, we were looking for anything."

In the Organs they found lead musket balls, pottery sherds, and the shafts of a few mines with ladders distinctive of the Spanish period. At night they crossed the old Cox ranch and the site of the Soledad mine. "The curious thing," said Newman, "was that they had military guards around the Soledad. That is why we had to

[1] The group consisted of Joe, Art Peevey, Gary Gibbons, Mike Rayburn, Russell Helterbrandt, Dick Nichols, Bill Batt, and later Kevin Henry.

cross there at night. Now I don't know why they are guarding an empty mine unless they are either taking something out of it or putting something into it."[2]

By the end of the sixth month, in August 1973, they had reached Hembrillo. At the time, the Bailey stories had broken out of Washington, and the Army had set up a patrol around Hembrillo. The Newman group noticed, however, that the patrol retired at night. "The moon was nearly full, so we decided to go in," said Joe.

"We found six entrances, three of which were blocked, three open. Two of these open shafts were on the saddle, and the other is the one on the top. The two in the saddle had mine tailings, material that had been taken out. The one on top, by far the oldest, had no tailings, and we could find none nearby. Inside, however, the walls of the shaft are sooty with smoke, and we figured it may have functioned as a chimney for smelting down below.

"We rappeled down the chimney, about 250 feet down. The whole chimney had remnants of very ancient timber. We found no evidence that anyone had been all the way to the bottom, except possibly Noss. We found wagon wheel spokes, which had been made into a winch, and we figured it could have been Noss's."

On that weekend, one of their members, Dick Nichols, was caught by the range riders and arrested. The others evaded discovery and managed to sneak out to a rendezvous point in the Jornada. There they were met by a driver in Newman's Jeep and taken back to El Paso.

Later, Joe Newman and twenty-one-year-old Kevin Henry returned. They made many trips. Henry, slight of build and agile, made most of the cave explorations. In one visit, he spent three days inside the peak.

"We found no skeletons, no river, nor anything that Noss described as seeing in the big cave," said Newman. "I think we either hit the place where Fiege went in or we found a brand-new place." Newman's voice dropped to a more guarded tone. "We found some bars.

"Once having found it, we didn't know what to do. The guys who had been caught didn't want to be caught a second time."

[2] The base spokesman at White Sands Missile Range denied to the author that guards had ever been posted at the Soledad.

The Army had warned them that a second offense would see them charged in federal court with trespassing.

"Sometime in early 1974, I was contacted by an acquaintance, who said a fellow named Fred Drolte wanted to meet with my group. We met, and Drolte wanted me to tell the size, shape, and look of each bar, to tell what else I'd seen, and where I'd gone in. I told him a little bit, not too much."

In early July 1974, Drolte signed Newman to a contract whereby Newman would show Drolte his discovery in return for 10 per cent of Drolte's overall share. Later that same month, Drolte asked Newman to go in and get some of the bars.

"Drolte had arranged for a helicopter to take me in at dusk and grab some bars. Drolte had this helicopter at the Las Cruces, and it was rigged with a net underneath, a net which was intended to haul gold.

"His plan had imagination, I'll give him that. When we went in, the pilot had to stay in the chopper to keep it running while I snaked into the cave and got out as many bars as I could. I was to throw the bars into the net, and as we went out we were to hover over one of the water tanks on Dick Hill's ranch, trip the net, and dump the bars in there.

"Fred said that if we got caught by the ranger riders I was to ask for a certain colonel and use Fred Drolte's name. Then we'd be released with no questions.

"Well, we tried it. We flew in with a net and everything. But we had caved in some of our tunnel intentionally, and there wasn't enough time for me to move the rocks and get back in there. We left with nothing."

A few days later, Newman, accompanied by Kevin Henry, returned to Hembrillo by his old route—up an arroyo by Jeep and by foot. "We stayed in there underground for three days. We got out with a bar, a full-sized bar. This was to be given to the governor by Bailey.

"I handed it over to Drolte and, later, Drolte cut off a pound piece and gave it to Bailey. The rest of the bar is still in El Paso. Drolte said it was originally planned to show the whole forty-pound bar to the governor, but the lawyers were against it. They said the Secret Service or the Army would confiscate it."

Fred Drolte, inaccessible in his guarded El Paso home and in his barricaded Caballo mine, has neither confirmed nor denied Joe Newman's story. Certain aspects, however, are corroborated. The Drolte-Newman contract exists; it was filed in federal court in Albuquerque. And *someone* did secretly enter Hembrillo in late July 1974. And, according to Department of the Army memos, there may have been military or Treasury Department collusion.

The pertinent memo is dated July 26, 1974, and is addressed to Norman Augustine, Assistant Secretary of the Army, and head of the Army's research and technical development.

Written by Augustine's aide, it states that Augustine had received an urgent call from Dr. Harold Agnew of the Atomic Energy Commission Laboratory at Los Alamos, New Mexico.

"Harold Agnew called," said aide K. C. Emerson. "I talked with him in your absence, to say that he was concerned about the gold problem at White Sands Missile Range.

"Earlier this week a group was allowed to go into the area for forty-eight hours.[3] They came out with a gold bar (they may also have taken it in). Harold has the bar and is analyzing it for age and content.

"This week he talked with Stark Draper to see if one of his instruments was sensitive enough to help. Harold said that if all the gold is present, as claimed in one heap, Stark can detect it as far away as one kilometer, and is interested.

"Harold said that because of all the problems with this governor of New Mexico, and other groups, that someone should be looking after the Army's interest. Now that Stark Draper has detectors that can help, maybe we [the Army] will want to do something."

The Army checked out the Draper instruments, which were at the Massachusetts Institute of Technology. MIT advised the Army "that their equipment is strictly for laboratory use and will not be available for field use for a few years. The MIT staff suggested that Professor Mueller at Ohio State did have a portable instrument which works on the same principle that would do the job. We checked with Professor Mueller's office and found to our surprise (and pleasure) that the instrument in question is now at the Army experimental laboratory."[4]

[3] Emphasis added by author.
[4] Department of the Army memo to K. C. Emerson from Marvin Lasser, director of Army Research (July 31, 1974).

The Army never used the Mueller instrument. Further memos pertaining to its use were deleted from the Army's response to a freedom-of-information request. Likewise deleted were the memos relating to the surreptitious entry of July 1974. Drolte had told Newman that there was protection. And someone, possibly Newman, had gone in. But the identity of the protection remains a government secret.

As for special detection equipment, the Army was not the only possessor of such instruments. Norman Scott, professional treasure hunter, had access to one and would play it as one of his lead cards in 1976.

XVI

Operation
Goldfinder

Mr. Scott contacted me in the legislative lobby. . . . We discussed the Noss treasure. He stated he wanted to have a look at the site stating that he got into this matter by being advised that Captain Fiege had cleared about 150 gold bars through a "gold free" port and had sold them. I told him that in order to get into the premises he would need permission from this office and the military. . . . He is going to continue his investigation. . . ."

—State land office memo
by Oscar Jordan
(March 7, 1973)

The significance of the memo is the date. Norman Scott and his brother Sam[1] were investigating the White Sands treasure a full month before F. Lee Bailey took his bar to Washington.

Norman Scott is president of Expeditions Unlimited, an archaeological recovery company owned by Texas financier Clint Murchison and headquartered in Pompano Beach, Florida. Expeditions Unlimited's accomplishments include excavation of the Mayan Sacred Well at Chichen Itza and exploration of Port

[1] It was Sam Scott who called on Oscar Jordan and is referred to in the Jordan memo.

Royal, the sixteenth-century Jamaican pirate capital, which slipped into the sea.[2]

Scott, born in 1929 and son and great-grandson of naval officers, was just getting into the treasure business in 1963 when he read of the museum excavation at White Sands.

"I had no access to the base," he recalled, "so I just tabled it. Our interest was revived late in 1972 when there was an article about Noss's widow, Mrs. Yancey. My brother Sam contacted her and we became her agent."[3]

In an ensuing four-year investigation, Scott demonstrated some brilliant detective work, including the uncovering of the identity and location of Willie Douthit.

After closing his field investigation in October 1976, Scott applied for an entry permit, stating:

"Our investigation shows that about 80 per cent of the representations of others from a period of 1937 through 1976 are not based on fact. They range in scope from generalizations to outright misrepresentations by persons for either profit or by being careless in their statements.

"Notwithstanding the above, there appears to be at least 20 per cent of the legends during the last four decades that cannot be proven false or misleading. In fact, there appears to be a preponderance of circumstantial evidence that indicates they may be factual.

[2] At Chichen Itza in Yucatan, Scott, his divers, and Mexican archaeologists retrieved bones of nearly five hundred sacrificial victims; gold jewelry and slippers; wooden furniture twelve hundred years old; fifteen hundred other artifacts, and two dolls. The latter, made of wood and wax, were recovered from a stratum that held thirteenth-century Mayan objects. The striking thing about the dolls was that they bore legible Latin inscriptions. Some scientists believe them to be evidence of European contact prior to Columbus. The Chichen excavation was a multimillion-dollar project that had defied archaeologists since the first attempts began in 1905. Scott agreed to solve the technical problems and to excavate at no cost to the Mexican Government, with all artifacts going to the government. In return he received media and promotion rights. He then sold book, magazine, TV, and newspaper rights and used those contracts as security for a loan to capitalize the venture. To reduce costs, he asked manufacturers to contribute equipment, personnel, and funds to the expedition in exchange for promotion accruing from the media exposure. By the time he had finished, he had put together allegedly the largest archaeological expedition in the Western Hemisphere.

[3] Interview with author (Oct. 1976).

"Expeditions Unlimited has spent in excess of $25,000 to date on the research, which we are prepared to substantiate as a matter of factual evidence. Our firm has interviewed over 325 individuals. . . .

"It is our conclusion that Victorio Peak was merely a subterfuge and that there is no treasure in said peak. But that the treasure, if in fact it does exist, is on other sections.

"Accordingly, it is respectfully suggested that the existing permit be amended to include the right of penetration in the confines of the overall permit.[4] This is at the sole expense of the contractor, and at no cost to the U. S. Government. It is the opinion of this writer that by allowing this amended method, the matter can be conclusively 'put to bed.' I need not belabor the point that you are aware of at least part of the activities of the Army and other governmental agencies involved in this project for over four decades, with no total conclusions reached on actual facts. . . .

"This amended permit request is based in part on the fact that Expeditions Unlimited has now secured the co-operation of all of the six registered claimants of record. This in itself, I might add, has been quite a venture, as it was one of the stumbling blocks in the past. As you know, the claimants were arguing among themselves as to the method of entry. . . .

"We will utilize the services of the Stanford Research Institute with their sophisticated electronic gear. We would like to have permission for six out of the ten days on site to utilize a dozer and front-end loader for the purpose of penetration in an area that the . . . Bailey group states that they know exactly where some of the material is. Since the Bailey group has now joined forces with us, we have looked into their statements and find some of them were based on fact. However, some were not, due to a lack of previous research on their part. . . .

"We have a list of the previous attempts (published and

[4] Scott had obtained an earlier permit in June 1976, to make a surface exploration of Victorio Peak with electronic instruments leased from the Stanford Research Institute. Under the Army terms, no digging would be allowed, but if the instruments detected a large amount of buried metal, then further negotiations would be held. If they showed no such deposits, then the claimants would abandon their entry efforts. The latter clause was opposed by F. Lee Bailey, and that opposition, coupled with some technical problems, caused Scott to postpone the entry.

nonpublished) of entries onto WSMR. Some of them were with and without the government's official knowledge. We now know who the various predominant groups consist and consisted of. . . .

"In the event that gold material is found we would obviously want a reasonable extension of a permit, in conformity with the convenience of the U. S. Government to extract said material and put it into the aforementioned impartial hands of a receiver. The purpose of the ten-day permit is to solely locate any material—not to extract same. . . .

"In the event that an economically producing mine is found, I believe that the same percentages would be applicable and, respectfully, consideration should be given to the operation of the mine for the furtherance of the recipients (that is, the Army Relief Fund, the state of New Mexico, the federal government, and the proper claimants).

". . . Our role at this particular point is to merely establish the credibility of the story and find the material, if in fact it exists. Or conversely, if it does not exist, to put the legend to rest, which is in the best interest of the U. S. Government—specifically the Department of the Army."[5]

While Washington officials of the Army considered Scott's letter, the White Sands command commenced a strange project. Bulldozers and other earth-moving equipment went onto Victorio Peak and caved in all known cave entrances. The entrances were then sealed with iron doors. Asked about this curious behavior— the sealing of caves just prior to the authorized exploration of those same caves—a White Sands spokesman explained it was "for safety reasons. We don't want anyone to fall into the holes and be hurt."

In early 1977, the Army authorized Operation Goldfinder. But even as Norman Scott began his preparations, he was expressing fear that the gold, if it had been there at all, was gone. The performance of the Secret Service and the White Sands command had not reassured him on the matter.

The skimpy investigations of illegal entries in November 1973 and July 1974 have already been discussed. An earlier entry was

[5] Entry application from Norman Scott to Bland West, deputy general counsel, Department of the Army (Oct. 22, 1976).

reported to the Treasury Department by F. Lee Bailey in May 1973. Said Bailey:

"I called Treasury's Ron Brooks and reported I had just been informed by my client that civilian men in two Jeeps and a pickup truck had been seen digging at one of our caches. They took quite a few bars and drove off. Brooks phoned me later and said, 'It is not *our* doing.' That was it. No further explanation, no further action. I was being burned by the government right in the middle of negotiations with the government."[6]

A year later, in the spring of 1974, House Appropriations Committee chairman Mahon of Texas, at the request of Roscoe Parr, asked the Army to investigate a report that Army officers in collaboration with CIA representatives were removing gold from White Sands and spiriting it into Mexico.

"This matter was investigated," Mahon was informed by the Army, ". . . and the report was found to have no factual foundation whatever. This matter is covered in separate documents which are available for inspection."[7]

One of the most persistent of the rumors is the "Zurich deal," first mentioned in a Secret Service report of September 26, 1973.

A Florida businessman had reported to the Service that he had been "in Boston on September 19, 1973, to sign a contract with an airline to fly the gold out of this country. He said that [name censored] of Tampa had arranged with Lloyds of London to insure the flight for a $232 million premium. . . . He also stated that he had learned that the Jeffries Bank of Brussels had been approached by [name censored] to finance the airline contract but had been unsuccessful. The airline involved is believed to be either TWA or Flying Tigers. . . ."

The agent writing the report believed that a potential fraud was involved and advised the informant that "this Service has no investigative jurisdiction of investment schemes involving gold until gold surfaces." He suggested the matter be referred to the U.S. attorney.

The rumor refused to die. Two years later, Norman Scott, after

6 Interview with author.
7 Department of the Army "Memorandum for the Record" dated and signed by Bland West (July 22, 1974). Records of the investigation itself have not been made available to the public.

interviewing several people, put together a never-proved story that gold was being smuggled out of White Sands in a truck. From there it went to a refinery at Chihuahua, Mexico, where it was given a "pedigree," a bill of sale saying it had been recently mined. The gold was then moved to Grand Cayman Island, site of fifty to sixty foreign banks, and thence to Zurich.

On March 8, 1976, Joe Newman called on Oscar Jordan at the state land office in Santa Fe. Newman had evidence of renewed Army digging and, as it developed, another Zurich rumor.

A Jordan memorandum dated March 9 stated that Newman "showed me two sets of pictures. One which was supposed to be taken in 1973 by [newspaperman] Howard Bryan and two recent pictures which were supposed to be taken by a party who they did not name.

"From the pictures it would appear that there had been some digging at the face of the mountain. It appeared that a deep trench had been dug at the side of the mountain. They [Newman and partner Dick Nichols] felt that someone had used heavy equipment to dig the trench. They wanted to know what could be done. . . ." Jordan suggested they complain to the U.S. attorney and "he might cause an investigation to be made. On the other hand, he might wish to prosecute whoever took the pictures. I suggested they check with their attorney before proceeding. . . .

"Newman and Nichols are fearful that somebody has already recovered the treasure and is shipping it out and they are torn between talking to the judge and the U.S. attorney or the alternative of keeping quiet and wait for the permit."

Before leaving, Newman told Jordan that one of the claimants, William Shriver, had heard rumors in Europe of a Zurich shipment.

Confirmation of a sort came ten days later. In a March 19, 1976, memorandum, Jordan reported that he had just finished a conference call with Norman and Sam Scott and two others.

"Sam reported that he had just talked to Shriver, who has returned from England and is in Florida and was advised that eleven tons of gold went out of White Sands Missile Range on March 10, 1976 by TWA and Swissair to Credit Swiss in Zurich, Switzerland; that it cleared in Zurich on March 10, shipped apparently out of Albuquerque International Airport, possibly through Chicago. Further, that there is additional material secreted in some lakes near El Paso on the Fort Bliss Military Reservation."

Jordan recommended "that perhaps they could get an investigation by the federal government which could go international and check out the shipments to Zurich. They could also make an on-the-spot investigation of the range itself (with military permission) and could take divers and check out the lake, all of which would be at the discretion of the FBI. . . ."

In a postscript closing out the memo, Jordan reported, "Norman Scott called and . . . stated that he personally had talked to Shriver and that Shriver assures him that he has a witness who is 'possibly a U.S. senator' who will furnish the names, dates, and places pertaining to the removal of the shipment; that if he is given immunity from criminal prosecution, he [the senator] will so testify. I suggested that Scott contact their attorney and have him contact the U.S. attorney in Albuquerque to secure the immunity."

Shriver, for some reason, was reluctant to approach the U.S. attorney. The witness referred to, however, was later tentatively identified as Ralph Yarborough, former U.S. senator from Texas. Yarborough was interviewed by veteran investigative reporter Hugh Aynsworth of the Dallas *Times-Herald*. "He convinced me," said Aynsworth, "that he knew nothing about it. I've known Yarborough a long time and I believe him."[8]

As for the digging at White Sands, the Washington *Post* of June 21, 1976, reported that the White Sands command firmly denied such digging. "There has been no heavy equipment" at the peak since 1963, said a spokesman.

The Army's own photographs, however, published in *Rolling Stone* magazine, show bulldozers and trucks on Victorio in 1973 and 1974.[9]

Operation Goldfinder began on March 18, 1977. The Army allowed ten working days of exploration and digging.

The claimants, the scientists, and the press entered Hembrillo

[8] Interview with author.

[9] The photos appear in the story "A Hundred Tons of Gold" (Dec. 18, 1975).

Basin in a long caravan of Jeeps, Broncos, Scouts, and four-wheel-drive trucks. In the lead vehicle was Norman Scott. Immediately following him, in a carpeted, air-conditioned Jeep, was Joe Newman and Kevin Henry. Elsewhere in the line strung out behind were Pat Patterson, Ova Noss, the Shrivers, Tom Jolley, and Leonard Fiege.

Fiege's cave was located on the second day. The explorer was Kevin Henry, and he came out to report that he had been in that same cave twice before—in 1974 and 1975. In those visits, he found no stacks of gold or bars of any kind. In the 1977 visit, he found that the interior had been entirely changed.

"The blasting," said Henry, "done by the Army since then has changed everything inside Victorio Peak. There were three caves I was in before and all are now bulldozed shut by the Army—dynamited shut and sealed by iron doors."

In 1961, Fiege had passed an Army-administered lie detector test that he saw gold bars stacked in a cave room. In 1974 and 1975 Kevin Henry entered the same cave room and found it empty. In 1977, the room was caved in and sealed with an iron door.

Fiege said the gold had been stolen. "I have my suspicions but I can't prove it. I think it was in the period after I took the test in 1961. They did a lot of mining here that nobody knew about."

Norman Scott was impressed by the enormous amount of mining evidence found within Victorio Peak. "It's more than is accounted for by the Army records," said Scott.

Pat Patterson's group attempted to excavate at the "cottonwoods" near the mural. They were stopped, however, by a state archaeologist, who claimed, on the basis of two bits of rock chips, that they were disturbing a valuable archaeological site. Patterson's group also was stymied at Geronimo Peak, where radar instruments of the Stanford Research Institute indicated the presence of a large cavern, possibly connected underground to Victorio Peak.

"We did not dig there," said Scott, "because Patterson's group could not show us an entrance. We would have had to mine our way into the mountain."

Harvey Snow was scheduled to show up and lead Norman Scott

to the alleged trove at Hard Scrabble. But Snow, fearful of something and claiming a sudden illness, backed out at the last moment.

Joe Newman took Scott to a sealed entrance, where he claimed to have found bars. Scott, however, placed a low priority on Newman's claim and never allowed the bulldozer to dig at the Newman hole.

Other, lesser, claimants proved fraudulent.

The claimant who proved the most credible was Ova Noss. In the 1930s, Doc Noss had drawn a map showing the main cavern, "large enough for a freight train," four hundred feet down in the bowels of Victorio Peak.

On the last day of Operation Goldfinder, the Standford radar mapped the deepest bowels of Victorio. "It showed," said Norman Scott, "a main cavern at least a hundred feet in diameter, and a hundred feet high. It may have been much larger prior to the Noss explosion. And there are passageways leading from it, suggesting cave rooms. The uncanny thing is that the configuration on the Stanford radar matches almost exactly the old Noss sketch."

In April 1977, the Stanford Research Institute released its written report on the radar surveys at Victorio Peak. The SRI concluded that Noss's physical description of the cave, and the cave-in, were verified.

"Radar investigation of the mountain reveals the existence of a very large cavern at the base of the 400 foot peak, near the center apparently associated with the junction of two crossing gash fissures and the east-west, vertical intrusive diorite dike that bisects Victorio Peak."

Examination of the original Noss summit shaft showed that (at 174 feet depth) a "large triangular wedge of loose mud and gravel has fallen in relatively recent times blocking further access along this fissure to the lower depths of the mountain." The report suggested further that Ova Noss, in her many attempts to re-excavate the shaft, had dug in the wrong direction. "Had the shaft been an incline to the west, it would have most probably penetrated to the large cavern.

In conclusion, SRI said enough verification of the Noss story had been gathered in Operation Goldfinder to warrant further exploration. "In view of the . . . so-far-proven consistency of the original Noss story a further exploration of the peak should be undertaken using microgravity, resistivity, a high-powered radar and excavation."[10]

The operation turned up no gold, and it winnowed out a number of fraudulent claimants. Fiege's story was made moot. He passed a lie detector test, and Kevin Henry found the same cave room empty years later. Either Fiege beat the lie detector in 1961, or someone removed the bars after that.

Harvey Snow's tale looks suspect. He had a chance to prove it and backed out.

Patterson, Bailey's main client, likewise had a chance to lead Scott directly to gold and failed to do it at the Geronimo Peak site. Part of Patterson's story, however, is still untested—the portion dealing with the "cottonwoods." That site was unfortunately shut down by a young state archaeologist who may have acted precipitously.

Tony Jolley's story remains untested. Scott did not provide sufficient metal-detecting equipment to examine the Jolley sites where Jolley once buried bars at night with Doc Noss. However, Jolley remains a very convincing man.

Joe Newman's story was likewise untested, due to unfriendly relations with Norman Scott.

Is there gold at Victorio Peak? Norman Scott was encouraged by several corroborations of the Noss story, and most especially encouraged by the matching up of the radar and Noss's sketch. "I think the Noss information is basically corroborated," said the cautious Scott. "What we now need is to go in for a full-scale mining operation; spend three hundred thousand dollars and two to three months to dig into the bowels of the peak."

Scott, the deep-sea explorer, stared with blue eyes at the high, desert mountain land around Victorio. It was the last day of the dig. No gold had been found.

"I am curious about all the mining evidence we found in the peak. There was a lot of work done by the Army here. A lot of work. Many months of activity.

[10] L. T. Dolphin, W. B. Beatty, and J. D. Tanzi, "Radar Probing of Victorio Peak, New Mexico" (Menlo Park, Calif.: Stanford Research Institute, 1977).

"There is something here. Many times I think I am skirting the periphery of something big."

Is there gold?

Operation Goldfinder did not resolve the question.

And there are other questions.

Why has the Secret Service lost, hidden, or destroyed the hundreds of documents relating to the White Sands, including the once-voluminous file on Doc and Ova Noss, the file that was made available to Colonel Gasiewicz in 1963?

Why has the White Sands command repeatedly misrepresented or hidden its activities, including the museum findings in 1963 and the illegal entries and excavations in 1973–76?

Why were the caves filled by bulldozers and sealed by iron doors just prior to Operation Goldfinder?

Finally, there is the mystery of Fred Drolte. In 1973, Drolte delivered a one-pound sample of gold bar to F. Lee Bailey for testing in Washington. In 1974, Drolte delivered a second one-pound bar to Bailey, which was tested at Los Alamos. (Both bars tested at approximately 60 per cent and indicated ancient smelting.) In 1976, Drolte staked a claim in the Caballos about fifty miles from Victorio Peak. He hired armed guards to protect the site.

One can envision a scenario whereby a person with the right connections makes secret entry into Hembrillo, removes the gold, and plants it in a legally claimed mining site. When matters cool down, he or she say they have struck gold and produce it for sale. Who can say it was taken from White Sands?

It sounds far-fetched, perhaps. But in 1977, Fred Drolte—the one man known to have produced gold bars—was the sole treasure hunter who did not show up at Operation Goldfinder. He told Joe Newman and Pat Patterson that he had no interest in the expedition.

Index

Capoques, the, 33
Carlotta, Empress, 51
Carriago, Tony, 104
Carrizozo Malpais, 24
Carson, Xanthus, 91, 92n, 158, 159
Castaneda, Pedro de, 30, 37
Ceuola, 39, 41
Chalchihuitl, Mount, 43
Chavez, Fray Angelico, 23n, 71
Cheyenne Mining Company, 121–22, 125, 127, 137
Chibcha Indians, 30–31
Chino, Wendell, 79
Chris, L. C., 54–55
CIA (Central Intelligence Agency), 2, 156, 187
Cibola, Seven Cities of, 29, 30–45, 47, 104, 127
Clippinger, Donn, 27, 29, 129
Clovis, N. Mex., 114, 126, 127
Cochise, Chief, 78, 82
Columbus, Christopher, 31, 184n
Copper mines, 48, 49
Coronado, Francisco (Coronado expeditions), 30, 37–38, 42–44, 104
Coronado's Children (Dobie), 50
Cortez, Hernando, 7, 30
Cottonwoods, the, 170, 192
Cox ranch, 55, 62, 64, 178
Coyotes, 24–25
Cravens, Jack, 107
Cruse, Thomas, 84–86

Daklugie, Ace, 78
Daught (Douthit), Willie, 8, 10–11, 12, 20, 53, 56, 88, 90–100, 103, 104, 108, 184; background, described, and gold find, 8, 10–11, 12, 90–100
Davis, L. H., 61–67
Dean, John, III, 2, 68, 70, 156, 165, 169
Doan, Judge, 96
Dobie, J. Frank, 25, 50–51, 54
Dona Ana County (N. Mex.), 46, 55, 56, 57, 60, 61–62, 91, 118, 123, 177
Dorantes, Andres, 32, 35, 36
Douthit, Willie. *See* Daught (Douthit), Willie

Draper, Stark, 181
Drolte, Fred, 156, 157, 158, 161, 163, 164, 167, 172–75, 176, 177–78, 180–81, 182, 193; background, described, 156, 173, 174
Dunham, Charles Kingsley, 66–67

Ehrlichman, John, 156, 161–63, 164, 165, 166–67
Eisenhower, Dwight D., 4–5
El Dorado myth, 30
Elephants (mammoths), 43, 127
El Paso, Tex., 53–54, 55, 64, 93, 94, 97, 99, 113, 122, 144, 156, 176–80 *passim*, 189
Emerson, K. C., 181
Estavanico, 32, 104
Esteban, 38–41, 42n
Expeditions Unlimited, 183–85

FBI (Federal Bureau of Investigation), 124, 138, 170, 171, 189
Fiege, Leonard V., 18–19, 56, 88, 140–45, 150, 152, 153, 154, 163, 165, 179, 183; and Goldfinder Operation, 190, 192; and Noss treasure find, 18–19, 56, 88, 140–45, 163
Fincher, Claude, 103, 104, 119, 121
Fisher, Mel, 7
Fluorspar (fluorite), 105
Ford, Gerald R., 79
Fountain, Albert J., 57–62, 63, 64, 67, 83
Franklin, Mount, 54–55
Franklin (now El Paso), Tex., 53–54. *See also* El Paso, Tex.

Gaddis Mining Company, 151–52, 153, 155
Garman, Ralph, 143
Garrett, Pat, 57–58, 61, 103, 127
Gasiewicz, Sigmund I., 142, 143, 146–47, 193
Gavilan, Captain de, 50
"Geology of the Organ Mountains," 66
Geronimo, Chief, 50–51, 82, 86
Geronimo Peak, 140, 164, 190
"Gilded Man, The," 30–31